Conversion Factors Between the United States Customary System and the Système International d'Unités

Quantity	To Convert from ...	To ...	Multiply by ...
Length	foot (ft)	meter (m)	0.3048
Length	inch (in.)	meter (m)	0.0254
Length	mile (mi)	kilometer (km)	1.609
Volume	gallon (gal)	meter^3 (m^3)	3.785×10^{-3}
Volume	gallon (gal)	liter (L)	3.785
Mass	slug	kilogram (kg)	14.59
Force	pound (lb)	newton (N)	4.448
Pressure or stress	pound/inch^2 (psi)	pascal (Pa)	6895
Work, energy, or heat	foot-pound (ft·lb)	joule (J)	1.356
Power	foot-pound/second (ft·lb/s)	watt (W)	1.356
Power	horsepower (hp)	kilowatt (kW)	0.7457

Quantity	To Convert from ...	To ...	Multiply by ...
Length	meter (m)	foot (ft)	3.281
Length	meter (m)	inch (in.)	39.37
Length	kilometer (km)	mile (mi)	0.6214
Volume	meter^3 (m^3)	gallon (gal)	264.2
Volume	liter (L)	gallon (gal)	0.2642
Mass	kilogram (kg)	slug	0.0685
Force	newton (N)	pound	0.2248
Pressure or stress	pascal (Pa)	pound/inch^2 (psi)	1.450×10^{-4}
Work, energy, or heat	joule (J)	foot-pound (ft·lb)	0.7376
Power	watt (W)	foot-pound/second (ft·lb/s)	0.7376
Power	kilowatt (kW)	horsepower (hp)	1.341

www.brookscole.com

brookscole.com is the World Wide Web site for Brooks/Cole and is your direct source to dozens of online resources.

At brookscole.com you can find out about supplements, demonstration software, and student resources. You can also send email to many of our authors and preview new publications and exciting new technologies.

An Introduction to Mechanical Engineering

Jonathan Wickert
Carnegie Mellon University

THOMSON

™

BROOKS/COLE

Australia • Canada • Mexico • Singapore • Spain • United Kingdom • United States

THOMSON
BROOKS/COLE

Publisher: *Bill Stenquist*
Editorial Assistants: *Valerie Boyajian, Julie Ruggiero*
Technology Project Manager: *Burke Taft*
Executive Marketing Manager: *Tom Ziolkowski*
Marketing Communications: *Margaret Parks*
Editorial Production Project Managers: *Mary Vezilich, Kelsey McGee*
Composition Buyer: *Ben Schroeter*
Print/Media Buyer: *Doreen Suruki*

Permissions Editor: *Kiely Sexton*
Production Service: *Martha Emry Production Services*
Copy Editor: *Pam Rockwell*
Illustration: *Atherton Customs*
Text and Cover Designer: *Roy R. Neuhaus*
Cover Image: *Courtesy of FANUC Robotics North America Inc.*
Cover Printing, Printing and Binding: *Phoenix Color Corp.*
Compositor: *Atlis*

For more information about our products, contact us at:
Thomson Learning Academic Resource Center
1-800-423-0563

For permission to use material from this text, contact us by:
Phone: 1-800-730-2214
Fax: 1-800-730-2215
Web: http://www.thomsonrights.com

Library of Congress Control Number: 2002117171

ISBN 0-534-39132-X

Brooks/Cole—Thomson Learning
10 Davis Drive
Belmont, CA 94002
USA

Asia
Thomson Learning
5 Shenton Way #01-01
UIC Building
Singapore 068808

Australia/New Zealand
Thomson Learning
102 Dodds Street
Southbank, Victoria 3006
Australia

Canada
Nelson
1120 Birchmount Road
Toronto, Ontario M1K 5G4
Canada

Europe/Middle East/Africa
Thomson Learning
High Holborn House
50/51 Bedford Row
London, WC1R 4LR
United Kingdom

Latin America
Thomson Learning
Seneca, 53
Colonia Polanco
11560 Mexico D.F.
Mexico

Spain/Portugal
Paraninfo
Calle Magallanes, 25
28015 Madrid
Spain

Contents in Brief

Contents

Student Preface

PURPOSE

This textbook will introduce you to the field of mechanical engineering and help you appreciate how engineers design the hardware that builds and improves our society. As the title implies, this textbook is neither an encyclopedia nor a comprehensive treatment of the discipline. Such a task is impossible for a single textbook, and, regardless, my perspective is that the traditional four-year engineering curriculum is just one of many steps taken during a lifelong education. By reading this textbook, you will discover the "forest" of mechanical engineering by examining a few of its big "trees," and along the way you will be exposed to interesting and practical elements of the profession called mechanical engineering.

APPROACH AND CONTENT

This textbook is intended for students who are in the first or second years of a typical college or university program in mechanical engineering or a closely related field. Throughout the following chapters, I have attempted to balance the treatments of problem-solving skills, design, engineering analysis, and practical technology. The presentation begins with a narrative description of mechanical engineers and what they do (Chapter 1), and an outline of good problem-solving skills, particularly with respect to numerical values and unit systems (Chapter 2). Six elements of mechanical engineering are emphasized subsequently in Chapter 3 (Machine Components and Tools), Chapter 4 (Forces in Structures and Fluids), Chapter 5 (Materials and Stresses), Chapter 6 (Thermal and Energy Systems), Chapter 7 (Motion of Machinery), and Chapter 8 (Mechanical Design). Some of the applications that you will encounter along the way include internal combustion engines, rapid prototyping, computer-aided engineering, robotics, magnetic resonance imaging, jet engines, automatic transmissions, and solar power.

What should you be able to learn from this textbook? First and foremost, you will discover who mechanical engineers are, what they do, and what technologies they create. Section 1.3 details a "top ten" list of the profession's achievements. Take a look at it now for a glimpse of mechanical engineering technologies, and recognize how the profession has improved your day-to-day life and society in general. Second, you will learn some of the order-of-magnitude approximations and back-of-the-envelope calculations that mechanical engineers can perform. To accomplish their jobs better and faster, mechanical engineers combine mathematics, science, computer-aided engineering tools, experience, and hands-on skills. Third, you will find that engineering is a practical endeavor with the objective of designing things that work, that are cost-effective to manufacture and sell, that are safe to use, and that are responsible in terms of their environmental impact.

You will not be an expert in mechanical engineering after having read this textbook, but that is not my intention, and it should not be yours. If my objective has been met, however, you will set in place a solid foundation of problem-solving, design, and analysis skills, and those just might form the basis for your own future contributions to the mechanical engineering profession.

Instructor Preface

APPROACH

This textbook is intended for a course that provides an introduction to mechanical engineering during either the freshman or sophomore years. Over the past decade or so, many colleges and universities have taken a fresh look at their engineering curricula with the objective of positioning engineering skills and applications earlier in their programs. Particularly for the freshman year, the formats vary widely, and content includes descriptions of "who are mechanical engineers" and "what do they do," seminars, design experience, problem-solving skills, basic analyses, and case studies. Courses at the sophomore level often emphasize design projects, exposure to computer-aided engineering, principles of engineering science, and a healthy dose of mechanical engineering hardware.

Core engineering science courses (for example, fluid mechanics, strength of materials, and dynamics) have evolved since the post-World War II era into their present relatively steady states. On the other hand, little standardization exists among introductory mechanical engineering courses at the freshman and sophomore levels. With limited discipline-specific instructional materials available for such courses, I believe that an important opportunity remains for attracting students, exciting them with a view of what to expect later in their program of study, and providing them with a foundation of good design, analysis, and problem-solving skills.

OBJECTIVES

While developing this textbook, my objective has been to provide a resource that others can draw upon when teaching introductory mechanical engineering to first- and second-year students. I expect that most such courses would encompass the bulk of material presented in Chapter 1 (The Mechanical Engineering Profession), Chapter 2 (Problem-Solving Skills), Chapter 3 (Machine Components and Tools), and Chapter 8

(Mechanical Design). The descriptions in Sections 6.6 through 6.8 of internal combustion engines, electrical power generation, and jet engines are largely expository in nature, and that material can be incorporated in case studies to demonstrate the operation of important mechanical engineering hardware. Based on the level and contact hours of their particular courses, instructors can select additional topics from Chapter 4 (Forces in Structures and Fluids), Chapter 5 (Materials and Stresses), Chapter 6 (Thermal and Energy Systems), and Chapter 7 (Motion of Machinery). For instance, Section 4.5 on buoyancy, drag, and lift forces is largely self-contained, and it provides an introductory-level student with an overview of some issues in fluid mechanics.

This textbook reflects my experiences and philosophy for introducing students to the vocabulary, skills, applications, and excitement of the mechanical engineering profession. My writing over the past six years has been motivated in part by teaching Fundamentals of Mechanical Engineering, a large course at Carnegie Mellon University that is open to both majors and nonmajors. This course includes lectures, a computer-aided design and manufacturing project, and a team design project (a portion of which is outlined in Section 8.3 in the context of design conceptualization). A number of vignettes and case studies are also discussed to demonstrate for students the realism of what they are learning: the Space Shuttle Challenger, the Kansas City Hyatt hotel, the "top ten" list of achievements developed by the American Society of Mechanical Engineers (Section 1.3), Air Canada Flight 143 (Section 2.2), the Mars Climate Orbiter spacecraft (Section 2.5), the twin towers of the World Trade Center (Examples 6.3, 6.7, and 6.9), integrated computer-aided engineering (Section 8.4), the first automobile automatic transmission (Section 8.5), and patents in engineering (Section 8.6).

CONTENT

I have not intended this textbook to be an exhaustive treatment of mechanical engineering, and I trust that it will not be read in that light. Quite the contrary: in teaching first- and second-year students, I am ever conscious of the mantra that "less really is more." To the extent possible, I have resisted the urge to add just one more section on a particular subject, and I have tried to keep the material manageable and engaging from the reader's perspective. Indeed, many topics that are important for mechanical engineers to know are simply not included here; this is done intentionally (or, admittedly, by my own oversight). I do have confidence, however, that students will be exposed to those otherwise omitted subjects in due course throughout the remainder of their engineering curricula.

I have not relied on any mathematics beyond algebra, geometry, and trigonometry, and there are no integrals or cross-products in this textbook. As with many general-purpose rules, I do admit one exception: Section 7.6 includes differentiation within the context of mechanisms and cams in internal combustion engines. That material is intended to motivate students during their concurrent study of mathematics through several concrete and easily visualized applications. Overall, this textbook's content should be readily accessible to any student having a conventional secondary school background in mathematics and physics.

In Chapters 3 through 8, I have selected a subset of mechanical engineering elements that can be sufficiently covered for early students to develop useful design, analysis, and problem-solving skills. The coverage has been chosen to facilitate the textbook's use within the constraints of courses having various formats. In particular, I have selected content that I have found to

- Match the background, maturity, and interests of students early in their study of engineering
- Help students think critically and learn good study and problem-solving skills, particularly with respect to order-of-magnitude approximation, double-checks, and the book-keeping of units
- Convey aspects of mechanical engineering science and empiricism that can be applied at the freshman and sophomore levels
- Expose students to a wide range of hardware and the hands-on nature of engineering
- Generate excitement through applications encompassing computer-aided design, medical imaging, aircraft, space flight, engines, automobile transmissions, ocean thermal energy conversion, nuclear power generation, and more

To the extent possible at the freshman and sophomore levels, the exposition, examples, and homework problems have been drawn from realistic applications. Because I find engineering to be a visual and graphical activity, I have placed particular emphasis on the quality and breadth of the nearly three hundred photographs and illustrations, many of which were provided by colleagues in industry, federal agencies, and academe. My view has been to leverage that realism and motivate students through interesting examples that offer a glimpse of what they will be able to study in later courses, and subsequently practice in their own careers.

Acknowledgments

It would have been impossible to develop this textbook without the assistance and contributions of many people and organizations, and, at the outset, I would like to express my appreciation to them. Foremost, generous support was provided by the Marsha and Philip Dowd Faculty Fellowship, which encourages educational initiatives in engineering. Adriana Moscatelli, Jared Schneider, Katie Minardo, and Stacy Mitchell, who are now alumni of Carnegie Mellon University, helped to get this project off the ground by drafting many of the illustrations that appear in the following chapters. The expert assistance provided by Ms. Jean Stiles in preparing and proofreading the textbook and the instructor's manual was indispensable, and I very much appreciate the many contributions she made.

My faculty colleagues and teaching assistants at Carnegie Mellon have provided many valuable comments and suggestions over the past six years. I am likewise indebted to my students in Fundamentals of Mechanical Engineering. Their collective interest and enthusiasm provided the forward momentum that was needed to bring this work to closure. Joe Elliot and John Wiss kindly offered engine dynamometer and cylinder pressure data to better frame the discussion of internal combustion engines in Chapter 6. Solutions to the homework problems were meticulously prepared by Brad Lisien and Al Costa, who are the lead authors of the companion instructor's manual. In addition, I gratefully acknowledge the feedback of the following reviewers: John R. Biddle, California State Polytechnic University at Pomona; Robert Hocken, University of North Carolina at Charlotte; Damir Juric, Georgia Institute of Technology; Pierre M. Larochelle, Florida Institute of Technology; Anthony Renshaw, Columbia University; and Gloria Starns, Iowa State University.

On all counts, I have enjoyed interacting with the editorial staff at Brooks/Cole— Thomson Learning. As the textbook's publisher, Bill Stenquist was committed to developing a high quality product, and it was a pleasure to collaborate with him. Martha Emry managed the textbook's production with skill and professionalism, and with a keen eye for detail. Other members of the team included production project manager Mary Vezilich, senior art director Vernon Boes, executive marketing manager Tom

Ziolkowski, editorial assistants Valerie Boyajian and Julie Ruggiero, and permissions editor Kiely Sexton. To each of them, I express my thanks for a job well done.

In conjunction with this textbook's design and analysis content, I have attempted to introduce readers to the hardware and applications of mechanical engineering. In that vein, colleagues at the following academic, industrial, and governmental organizations were remarkably helpful and patient in providing photographs, illustrations, and technical information: General Motors, Intel, Fluent, General Electric, Enron Wind, Boston Gear, Mechanical Dynamics, Caterpillar, NASA, NASA's Glenn Research Center, W. M. Berg, FANUC Robotics, Bureau of Reclamation, Niagara Gear, Velocity11, Stratasys, Carnegie Mellon, National Robotics Engineering Consortium, Lockheed-Martin, Algor, MTS Systems, Westinghouse Electric, and Timken. In particular, Sam Dedola and John Haury of Medrad, Incorporated, went the extra mile and developed numerous illustrations for the discussion of computer-aided design in Section 8.4. I've surely not listed everyone who has helped me with this endeavor, and I apologize for any inadvertent omissions that I may have made.

Last, but in no means least, let me express heartfelt gratitude to Karen and Becky for their patience and encouragement throughout this project.

Jonathan Wickert

About the Author

A Professor of Mechanical Engineering at Carnegie Mellon University, Jonathan Wickert teaches and conducts research in the areas of applied mechanics, dynamics, and mechanical vibration. As a researcher and consultant, he has worked with companies and federal agencies on a diverse range of engineering problems including computer disk and tape drives; the manufacture of sheet metal, fiberglass, polymers, and industrial chemicals; automotive brakes; radial flow gas turbines; and various consumer products. Dr. Wickert received his B.S., M.S., and Ph.D. degrees in mechanical engineering from the University of California, Berkeley, and he was a postdoctoral fellow at the University of Cambridge. He has served as associate editor of engineering journals, as a division chair in the American Society of Mechanical Engineers, and as chair of the undergraduate mechanical engineering program at Carnegie Mellon. Dr. Wickert has received awards in recognition of his teaching and research from the Society of Automotive Engineers, and the American Society for Engineering Education, and he was elected a fellow of the American Society of Mechanical Engineers.

1 The Mechanical Engineering Profession

1.1 OVERVIEW

Engineering is the practical endeavor in which the tools of mathematics and science are applied to develop working and cost-effective solutions to the technical problems facing our society. Engineers design many of the consumer products that you use every day and a large number of other products that you do not necessarily see but which are used in business and industrial settings. Engineers also design the machinery that is needed to manufacture those products, the factories that produce them, and the quality control systems that guarantee the product's safety and performance. Engineering is all about making useful things that work.

Machine components, forces, materials, energy, motion, design
‑‑‑‑‑‑‑‑‑‑‑‑‑‑‑‑→

The discipline of mechanical engineering is concerned in part with the following "elements": machine components, forces, materials, energy, motion, and design. Mechanical engineers develop hardware that exploits those elements in order to serve a useful purpose. Original design is always the underlying factor: An engineer creates a machine, product, or system that helps someone to solve a technical problem (see Figure 1.1). The engineer might start from a blank sheet of paper, conceive something new, develop and refine it so that it works reliably, and all the while satisfy constraints of manufacturability, safety, and cost.

Internal combustion engines, robotic welding equipment, computer hard disk drives, prosthetic limbs, automobiles, aircraft, jet engines, and electric power generation plants are some of the thousands of technologies with which mechanical engineers work every day. It would not be much of an exaggeration to claim that for every product you can think of, a mechanical engineer was involved at some point in its design, materials selection, temperature control, quality assurance, or production. Even if a mechanical engineer didn't conceive or design the product per se, it's still a good bet that a mechanical engineer designed the machines that were needed to build, test, or deliver the product.

FIGURE 1.1   Mechanical engineers design and test complex high-speed machinery, such as this diesel power plant for highway construction equipment.
Source: Reprinted courtesy of Caterpillar Inc.

Mechanical engineers devise machines that produce or consume power over the remarkably wide range indicated in Figure 1.2—from milliwatts (mW) to gigawatts (GW). Few professions require a person to deal with physical quantities across so many orders of magnitude (one trillionfold, or a factor of 1,000,000,000,000), but mechanical engineering does. As an example at the lower end of the power range,

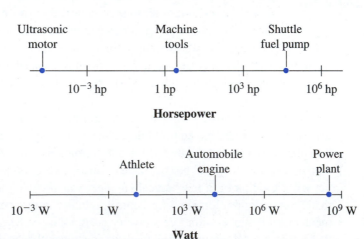

FIGURE 1.2   Mechanical engineers work with machines that produce or consume power over a remarkably wide range.

small precision ultrasonic motors, such as those used in a camera's autofocus lens, produce approximately 0.02 watts (W) of mechanical power. Moving upward in power level, an athlete using exercise equipment such as a rowing machine or stair climber can produce several hundred watts (about one-quarter horsepower) over an extended period of time. That output would be enough to power a household light-bulb. The electric motor in an industrial drill press might develop 1000 W (or 1 kW), and the engine on a sport utility vehicle produces about 100 times that amount. Near-ing the upper end of the power scale, the high-pressure fuel turbopump for the Space Shuttle's main engines (not the engines themselves, mind you, just the fuel pump) develops an output of 5.4×10^7 W, which is about 73,000 horsepower (Figure 1.3). Finally, a large commercial electrical power plant can generate 1 billion watts of power, an amount sufficient to supply a community of 800,000 households with electricity. Mechanical engineers design the fuel systems, steam generators, turbines, condensers, pumps, and pollution control systems that are expected to operate safely and around the clock in power plants.

■ **FIGURE 1.3**   Close-up view of a Space Shuttle main engine during a test in which it is swiveled to evaluate steering performance during simulated flight conditions.
Source: Reprinted with permission of NASA.

In this introductory chapter, we describe who mechanical engineers are, what they do, why they are dedicated to their profession, and what their notable accomplishments have been. After completing this chapter, you should be able to:

- Describe the differences between engineers and scientists and mathematicians.
- Describe the type of work that mechanical engineers do, and list some of the technical issues that they address.
- Discuss the types of industries and governmental agencies that employ mechanical engineers.
- List some of the products, processes, and hardware with which mechanical engineers work.
- Describe notable achievements of the mechanical engineering profession.
- Understand the structure and objectives of a typical curriculum for mechanical engineering students.

1.2   WHAT IS ENGINEERING?

The word *engineering* derives from the Latin root *ingeniere*, meaning to design or to devise, which also forms the basis of the word *ingenious*. Those meanings are quite appropriate summaries of the traits of a good engineer. Engineers apply their knowledge of mathematics, science, materials, and physical principles—as well as their skills in communications and business—in order to develop new and better products and machines. Rather than experiment solely through trial and error, an engineer designs faster, more accurately, and more economically by using mathematics, computers, and science as resources. In that sense, the work of an engineer differs from that of a scientist, who would normally emphasize the fundamental discovery of physical laws rather than their application to product development. Engineering serves as the bridge between scientific discovery, commercial applications, and business marketing (Figure 1.4). Engineering as a discipline does not exist for the sake of exploring mathematics, computation, science, or hardware by themselves. Rather, it is both a driver of social growth and an integral part of the economic business cycle, as we will discuss later in this chapter.

Many students begin to study engineering because they are attracted to the fields of mathematics and science. Others migrate toward engineering careers because they are motivated by an interest in technology and how everyday things work or, perhaps with more enthusiasm, in how not-so-everyday things work. Either way, it is important to recognize that engineering is distinct from the fundamental subjects of mathematics and science. At the end of the day, the objective of an engineer is to have built a device that performs a task that couldn't previously have been completed, or that could not have been completed so accurately, quickly, or safely. Mathematics and science are some of the tools that enable an engineer to test fewer mock-ups by refining designs

■ **FIGURE 1.4** A remarkable engineering achievement, the Space Shuttle Atlantis thunders into orbit as its twin solid rocket boosters and three main engines produce 7 million pounds of thrust.
Source: Reprinted with permission of NASA.

on paper and through computer models before any metal is cut or hardware is built (Figure 1.5).

Approximately 1.5 million people are employed as engineers in the United States. The vast majority work in industry; only about 12% are employed by federal, state, and local governments. Engineers who are federal employees are often associated with such organizations as the National Aeronautics and Space Administration, or the Departments of Defense, Transportation, and Energy. Only 3–4% of engineers are self-employed, and they work mostly in consulting or entrepreneurial capacities.

Traditional disciplines
----------------->
Engineering is generally broken down into five traditional core disciplines: mechanical, electrical, materials, civil, and chemical engineering.

In the next section, we will discuss mechanical engineering in depth, but let's first describe some of the other engineering fields. Electrical engineers

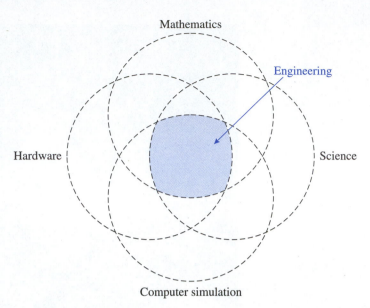

■ FIGURE 1.5   Engineers combine their skills in mathematics, science, computers, and hardware.

work in such areas as power transmission and distribution, integrated circuit design and production, networks and wireless communications, aviation electronics, and robot control systems. Materials engineers develop new compositions of matter that are used in applications encompassing semiconductors, lasers, steel and aluminum alloys, and magnetic media in computer hard drives. Among other activities, civil engineers design and construct roads, buildings, airports, tunnels, dams, bridges, and water supply systems. Chemical engineers are responsible for the industrial-scale production of chemicals and for managing the by-products in such fields as petroleum, electronics, and biotechnology. Even within all of these disciplines, there are dozens of other specializations and areas of emphasis. Figure 1.6 depicts employment statistics and the distribution of engineers in the five traditional disciplines as well as several others.

Engineers develop their skills through formal study in an accredited four-year college or university educational program and through practical work experience under the supervision of accomplished and more senior engineers. When starting a new design, engineers also use to advantage their common sense, physical intuition, hands-on skills, and judgment gained through previous technical projects. Engineers make approximate but useful back-of-the-envelope calculations in order to answer such questions as "Will a 10 hp engine be powerful enough to drive that air compressor?" or "How many *g*'s of acceleration must the blade in the turbocharger withstand?" When the answer to a particular question isn't known, or when more information is needed to complete a task, an engineer will conduct additional research to solve the problem, using such resources as books, professional journals,

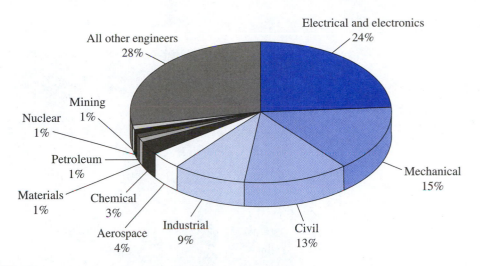

FIGURE 1.6   Percentages of engineers working in various engineering fields.
Source: United States Department of Labor.

and trade publications in a technical library; engineering conferences and product expositions; patents; and data provided by industrial vendors. The process of becoming a good engineer is really a lifelong endeavor and is a composite of both education and experience. Engineers are constantly learning new approaches and problem-solving techniques and informing others of their discoveries, particularly as technology, markets, and economies grow and evolve at a fast pace.

1.3   WHO ARE MECHANICAL ENGINEERS?

The field of mechanical engineering is concerned with machine components, the properties of forces, materials, energy, and motion, and with the application of those elements in order to devise new machines and products that improve society and people's lives (Figures 1.7, 1.8). The U.S. Department of Labor has described the profession as follows:

> Mechanical engineers research, develop, design, manufacture and test tools, engines, machines, and other mechanical devices. They work on power-producing machines such as electricity-producing generators, internal combustion engines, steam and gas turbines, and jet and rocket engines. They also develop power-using machines such as refrigeration and air-conditioning equipment, robots used in manufacturing, machine tools, materials handling systems, and industrial production equipment.

Mechanical engineers are known for working on a wide range of products and machines. Some examples are automobiles (Figure 1.9), aircraft, jet engines, computer hard drives, microelectromechanical acceleration sensors (used in automobile air bags), heating, ventilation, and air-conditioning systems, heavy construction equipment, cell phones,

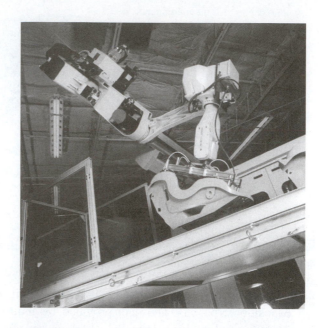

■ **FIGURE 1.7** Robots are extensively used in automated industrial assembly lines and in manufacturing applications that require precision when repetitive tasks are being performed, as in the assembly and arc welding operations shown here.

Source: Reprinted with permission of FANUC Robotics North America Inc.

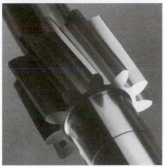

■ **FIGURE 1.8** Mechanical engineers design machinery and power transmission equipment using various types of gears as building block components.
Source: Reprinted with permission of Niagara Gear Corporation, Boston Gear Corporation, and W. M. Berg, Incorporated.

artificial hip implants, robotic manufacturing systems, replacement heart valves, planetary exploration and communications spacecraft, deep-sea research vessels, and equipment for detecting explosives.

Mechanical engineering is the second-largest field among the traditional engineering disciplines and is perhaps the most general. In 1998, approximately 220,000 people were employed as mechanical engineers in the United States, representing 15% of all engineers. The discipline is closely related to the technical areas of industrial (126,000), aerospace (53,000), petroleum (12,000), and nuclear (12,000) engineering, each of which evolved historically as a branch or spin-off of mechanical engineering. Together, mechanical, industrial, and aerospace engineering account for about 28% of all engineers. Other specializations that are frequently encountered in mechanical engineering include automotive, design, and manufacturing engineering. Mechanical engineering is often regarded as the broadest and most flexible of the traditional engineering disciplines, but there are many opportunities for specialization within a certain industry or technology that is of particular interest. Within the aviation industry, for example, an engineer will further focus on a single core technology, perhaps jet engine propulsion or flight control systems.

Above all else, a mechanical engineer makes hardware that works. An engineer's contribution is ultimately judged based on whether the product functions as it should.

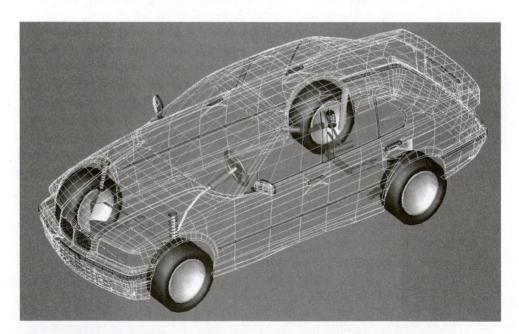

■ **FIGURE 1.9**   On a day-to-day basis, mechanical engineers use sophisticated computer-aided engineering tools to design, visualize, simulate, and improve products such as the automobiles shown here.

Source: Reprinted with permission of Mechanical Dynamics, Incorporated.

Mechanical engineers design equipment, it is produced by companies, and it is then sold to the public or to industrial customers in order to solve a problem and make an improvement. In the process of that business cycle, some aspect of the customer's life is improved and society as a whole benefits from the technical advances and additional opportunities that are offered by engineering research and development.

Mechanical engineering's top ten

- - - - - - - - - - - - - ▶

Mechanical engineering is not all about numbers, calculations, computers, gears, and grease. At its heart, the profession is driven by the desire to advance society through technology. The American Society of Mechanical Engineers (ASME) surveyed its members at the turn of the millennium in order to identify mechanical engineering's major achievements. (The society is the primary organization representing and serving the mechanical engineering profession in the United States and internationally.) This "top-ten" list, summarized here, will be useful for you in (1) better understanding who mechanical engineers are and (2) developing an appreciation of the contributions that mechanical and other engineers have made to your world. In descending order of the accomplishment's perceived impact on society, the milestones recognized in the survey are the following.

1. **The automobile.** The development and commercialization of the automobile were judged by mechanical engineers as the profession's most significant achievement in the twentieth century. Two factors responsible for the growth of automotive technology have been high-power lightweight engines and efficient mass-manufacturing processes. German engineer Nicolaus Otto is credited with designing the first practical four-cycle internal combustion engine, and after untold effort by engineers, it is today the power source of choice for most automobiles. In addition to improvements in the power plant itself, competition in the automotive market has lead to other improvements in the areas of automotive safety, fuel economy, comfort, and emissions control. Some examples are antilock brakes, run-flat tires, air bags, composite materials, computer control of fuel injection systems, satellite-based navigation systems, variable valve timing, and such alternative power technologies as electricity, biofuels, liquefied gas, and fuel cells.

The ASME recognized not only the development of the automobile but also the manufacturing technology behind it. Millions of vehicles have been produced inexpensively enough that the average family can afford one. Quite aside from his efforts at vehicle engineering, Henry Ford pioneered the techniques of assembly-line mass production, which enabled consumers from across the economic spectrum to purchase and own automobiles. The automotive industry, in turn, has grown to become a key component of the world economy, and it has also created jobs in the machine tool and raw materials industries. Automobile production is the largest manufacturing industry in the United States; it had some 1 million employees and $345 billion of economic output in 1999. No other technical field creates as much retail business or offers as many employment opportunities as the automotive industry. From minivans to stock car racing to Saturday night cruising, the automobile—one of the key contributions of mechanical engineering—has had a ubiquitous influence on our society and culture (Figure 1.10).

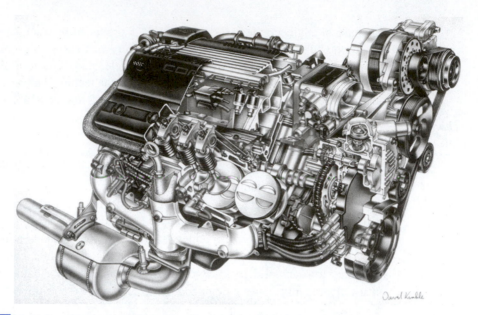

■ **FIGURE 1.10**   Mechanical engineers design, test, and manufacture internal combustion engines such as this eight-cylinder 5.7 liter engine.
Source: 2001 General Motors Corporation. Used with permission of GM Media Archives.

2. **The Apollo program.** In 1961, President John Kennedy challenged the citizens of the United States to land a man on the Moon and return him safely to Earth. The first portion of that objective was realized less than ten years later with the July 20, 1969 landing of Apollo 11 on the lunar surface. The three-man crew of Neil Armstrong, Michael Collins, and Buzz Aldrin returned safely several days later. The technological advances and profound cultural impact of the Apollo program caused it to be chosen as the mechanical engineering profession's second most influential achievement of the twentieth century.

The Apollo program was based on three key engineering developments: the huge three-stage Saturn V launch vehicle that produced some 7.5 million pounds of thrust at liftoff, the command and service module, and the lunar excursion module, which was the first vehicle ever designed to be flown only in space. Try to place the rapid pace of Apollo's development in perspective. Humans had dreamed of flight for thousands of years, but that dream was not realized until 1903, when Wilbur and Orville Wright made the first controlled and powered flight. They traveled 120 feet, a distance that is approximately one-third the length of the Saturn V launch vehicle. Only 66 years after that flight, the first lunar landing was witnessed live on television by millions of people around the world. The Apollo program is perhaps unique among engineering achievements in the way that it combined technological advances with the spirit of exploration and of patriotism. Indeed, photographs of Earth taken from the perspective of space have changed the way that we view ourselves and our planet. Apollo, space travel, planetary exploration, and even weather observation and communications

FIGURE 1.11   Astronaut John Young, commander of the Apollo 16 mission, leaps from the lunar surface at the Descartes landing site as he salutes the United States flag. The roving vehicle is parked beside the lunar module.
Source: Reprinted with permission of NASA.

satellites would all have been impossible without the initiative and dedicated effort of thousands of mechanical engineers (Figures 1.11, 1.12).

   3. **Power generation.** Mechanical engineers convert energy from one form to another, and they also move energy from one location to another. Abundant and inexpensive energy is an important factor behind economic growth and prosperity, and the distribution of electrical energy has improved the lives and standard of living for billions of people across the globe. In the twentieth century, economies and societies changed significantly as electricity was produced and routed to homes, businesses, and factories. Try to imagine the many ways in which electrical power contributes to your day-to-day activities. What would your life be like if you couldn't depend on electricity being available where and when you needed it? Indeed, we often take electrical power for granted, and power outages are viewed as being uncommon, surprising, and merely the result of something having gone wrong.

   Mechanical engineers are credited with developing the technologies to efficiently convert stored energy into electricity, which is more easily distributed and used than stored energy. Mechanical engineers manipulate chemical energy (coal, natural gas, and oil fuels, for instance); the kinetic energy of wind, which powers electricity-producing turbines; nuclear energy, which powers not only electrical plants but also ships, submarines, and spacecraft; and the potential energy of water reservoirs, which

FIGURE 1.12   Artist's concept of the International Space Station.
Source: Reprinted with permission of NASA.

feed hydroelectric power plants. Hydropower resources were developed primarily in the 1950s and most nuclear power plants were constructed in the 1960s; the many utility plants built in the United States since 1980 are fueled by cleaner-burning natural gas. Some issues that factor into electricity generation are the cost of the fuel, the cost of constructing the power plant itself, environmental impact, reliability, and safety. Large-scale power production is a prime example of the need for engineers to achieve a balance between technology and environmental and economic considerations. As natural resources are further consumed, or as they become more expensive either in terms of cost or pollution, mechanical engineers will be even more involved in developing advanced power-generation technologies, including solar, ocean, and wind power systems (Figures 1.13, 1.14).

4. **Agricultural mechanization.** In the early 1900s, an American farmer using the agricultural tools and methods of the day was able to produce enough food for only a handful of people. At the beginning of the twenty-first century, using more efficient technology, a farmer can now produce enough food to feed nearly 130 people. That 40-fold increase in productivity is due in large part to the abandonment of animal power in favor of mechanization.

The automation of farm equipment began with powered tractors in 1916 and with the introduction of the combine, which simplified grain harvesting. Other advances since that time have included improved weather observation and prediction, high-capacity

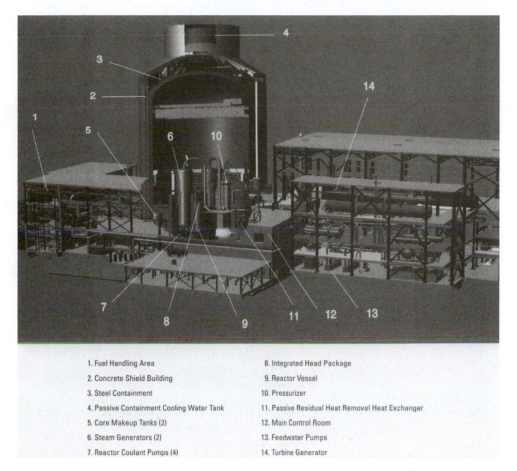

1. Fuel Handling Area	8. Integrated Head Package
2. Concrete Shield Building	9. Reactor Vessel
3. Steel Containment	10. Pressurizer
4. Passive Containment Cooling Water Tank	11. Passive Residual Heat Removal Heat Exchanger
5. Core Makeup Tanks (2)	12. Main Control Room
6. Steam Generators (2)	13. Feedwater Pumps
7. Reactor Coolant Pumps (4)	14. Turbine Generator

■ **FIGURE 1.13** The design for an advanced nuclear utility plant capable of producing 1 billion watts of electrical power.
Source: Reprinted with permission of Westinghouse Electric Company.

irrigation pumps, automated milking machines, and computer databases for the management of crops and the control of pests. With the advent of automation, more people have been able to take advantage of intellectual and employment opportunities in sectors of the economy other than agriculture. That migration of human resources away from agriculture has in turn helped to promote advances across a broad range of other professions and industries (Figure 1.15).

5. **The airplane.** Commercial passenger aviation has created domestic and international travel opportunities for both business and recreational purposes. The advent of air travel has enabled geographically scattered families to visit one another frequently, even at the spur of the moment. Business transactions are conducted in face-to-face meetings between companies located on opposite sides of a country or halfway across the world. In the past, five to six months were required to

■ **FIGURE 1.14** Mechanical engineers design renewable energy systems. The wind farm shown here incorporates several 1-million-watt turbines.
Source: Reprinted with permission of Enron Wind.

cross North America by ox team; two months were required to make the journey by steamboat and stagecoach; the steam train took four days; but the journey today takes six hours by commercial jet and is safer than it has ever been.

Mechanical engineers have contributed to nearly every stage of aircraft research and development and have brought powered flight to its present advanced state. One of their main contributions has been in the area of propulsion. Early planes were powered by piston-driven internal combustion engines, such as the Wright's 12-horsepower engine. When the jet engine was introduced, the cost of flying was reduced, because, for a certain amount of fuel, a jet-powered plane can carry a greater payload of cargo and passengers compared to a propeller-driven craft.

Some modern engineering accomplishments include the General Electric GE-90 engines that power some Boeing 777 jetliners and can develop a maximum thrust of over 100,000 pounds and the vectored turbofan engines used in military aircraft that enable the pilot to redirect the engine's thrust for vertical takeoffs and landings. Mechanical engineers are also responsible for designing the compressors, combustion systems, turbines, and control systems of advanced jet engines. They have spearheaded

■ **FIGURE 1.15**   Robotic vehicles under development will be able to learn the shape and terrain of a field of grain and harvest it with essentially no human supervision.
Source: Reprinted with permission of the National Robotics Engineering Consortium.

the development of aerospace-grade materials such as lightweight titanium alloys and graphite fiber-reinforced epoxy composites and of electromechanical "fly-by-wire" systems that adjust the airplane's aerodynamic surfaces to provide smooth, comfortable, and stable flight (Figures 1.16, 1.17).

6. **Integrated circuit mass production.** An amazing achievement of the electronics industry has been the miniaturization and mass production of integrated circuits, computer memory chips, and microprocessors. The vintage 8008 processor introduced by the Intel Corporation in 1972 had 2500 transistors, the 386^{TM} processor of the 1980s had 275,000 transistors, the Pentium® 4 microprocessor introduced in 2001 contained over 40 million transistors (Figure 1.18). Million-transistor integrated circuits are commonplace today. This remarkable rate of growth in the number of circuit components assembled on silicon chips is often called Moore's law, after Intel cofounder Gordon Moore. This observation states that, based on developments to date, the number of transistors placed on integrated circuits can be expected to double every 18 months. That prediction was made in 1965, and it still holds true, although engineers and scientists are now pushing up against fundamental physical limits for the density of circuit components.

An important process in producing semiconductor devices is lithography, in which complex circuit patterns are miniaturized and transferred from a pattern (called the mask) onto a silicon wafer. Transistors, capacitors, resistors, and wires are built as

■ **FIGURE 1.16** The airframe of the F-22 Raptor fighter is fabricated from aluminum, titanium, composites, and other advanced materials in order to meet the performance demands of low weight and high strength.

Source: Reprinted with permission of Lockheed-Martin.

small as possible in order to fit more memory or computing power into a given space. Conducting wires, for instance, are as narrow as 120 billionths of a meter (120 nanometers, or 120 nm) in high-capacity dynamic random access memory chips. To place that miniscule size into perspective, the diameter of a human hair is about 1000 times wider than those conductor wires (Figure 1.19).

Mechanical engineers work on the machines that manufacture those integrated circuits. They help design the machinery, alignment systems, temperature control, motors, bearings, and vibration isolation that enable integrated circuits to be so small and precise. The need to reproduce consistently microscopic circuit features has driven engineers to develop special krypton–fluorine laser illumination systems and equipment that can align entire masks and silicon wafers to a degree of precision that is within 70 nm (70×10^{-9} m). The level of accuracy that engineers have reached in massproducing integrated circuits is remarkable when you consider that the 70 nm tolerance is attained over the 300 mm diameter of a silicon wafer. The ratio of alignment accuracy to wafer diameter is 0.23 parts per million, which is equivalent to a gap of only 0.015 inches (a little more than the thickness of a business card) when viewed from a distance of one mile.

Engineers design structures that range in size from buildings, to aircraft, to automobiles, to hand tools, to items of microscopic dimensions. So-called micromechanical

■ **FIGURE 1.17**   The X-38 research vehicle is released from the wing pylon of its mother ship during a flight test. Mechanical engineers contribute to the technologies needed for an emergency crew return vehicle—or lifeboat—for use on an orbiting space station and based on the X-38.

Source: Reprinted with permission of NASA.

systems are fabricated by using the same clean-room techniques as are used in the production of integrated circuits, but the final result is a mechanism or mechanical structure instead of an electrical circuit. As an example, Figure 1.19 shows a device made up of a perforated plate suspended by four crableg flexures. Electrical wiring is routed to the comb fingers on the left and right sides, which in turn act as force actuators to move the mechanism back and forth.

7. **Air-Conditioning and refrigeration.** Over a relatively short period of time, the air-conditioning and refrigeration systems developed by mechanical engineers have made a significant improvement to our quality of life. The mechanical engineering achievement rated seventh is taken mostly for granted, but it is based on the physical principles of fluid mechanics, heat transfer, and thermodynamics and on the technologies of compressors, refrigerants, and heat exchangers.

Excessive heat has always been a source of problems related to reduced productivity in the summer, food spoilage, and even life-threatening medical conditions. Refrigeration systems are able to preserve and store food at its source, during transportation, and in the home. They enable seasonal produce to be stored fresh and allow consumers to forgo purchasing dairy products and meat on a daily basis out of concern that they would otherwise spoil. Even though food may be produced or grown thousands of miles

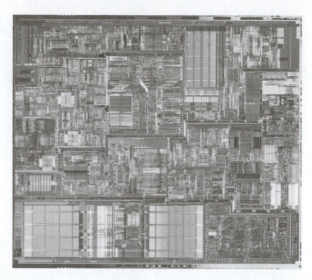

■ **FIGURE 1.18**   Mechanical engineers are involved in the production of integrated circuits and microprocessors and in the design of the lithography machines that make such devices as the Pentium® 4 microprocessor shown here.
Source: Reprinted with permission of Intel Corporation.

■ **FIGURE 1.19**   Mechanical engineers develop sensors, actuators, and mechanical structures out of silicon and other materials by using techniques originally developed for integrated circuits. The entire comblike structure shown here is smaller than the diameter of a human hair.
Source: Reprinted with permission of G. Fedder, Carnegie Mellon University.

away from the consumer, even in a different country, they can purchase it fresh any time of the day or night. Mechanical engineers have developed the technology behind that remarkable convenience and contribution to public health.

Although mechanical refrigeration systems were available as early as the 1880s, their application was limited to commercial breweries, meatpacking houses, ice-making plants, and the dairy industry. The general public had no way to store its own perishable food for an extended period of time. Early refrigeration systems required significant amounts of maintenance, and they were also prone to leaking hazardous or flammable chemicals—such as sulfur dioxide, methyl chloride, ethyl chloride, and isobutene—rendering them inappropriate for use in the home. The development in 1930 of the refrigerant Freon was a major turning point in the commercialization of safe residential refrigeration and air-conditioning. Since then, Freon has largely been supplanted by compounds that do not contain chlorofluorocarbons, which are now known to degrade the Earth's protective ozone layer.

8. **Computer-aided engineering technology.** The term *computer-aided engineering* (CAE) refers to a wide range of automation applications in mechanical engineering and other disciplines. CAE encompasses the use of computers for performing calculations, preparing technical drawings, analyzing designs, simulating performance, and even controlling machine tools in a factory. Mechanical engineers don't necessarily design the architecture of a computer, but they do use computers on a day-to-day basis. In fact, over the past several decades, computing and information technologies have changed the manner in which mechanical engineering is practiced. Most engineers, even those at smaller companies, have access to advanced computer-aided design and analysis software, information databases, and computer-controlled prototyping equipment. In some industries, these CAE technologies have entirely replaced traditional paper-based design and analysis methods.

Computer-aided engineering
- - - - - - - - - - - - - - - →

Developments in mechanical design automation began during the late 1950s, primarily for aerospace and defense-related systems. At that time, computer hardware, graphics technologies, and programming were all at a relatively low level when compared to today's capabilities. Because of the expense involved, most early applications concentrated on computer drafting. However, the costs involved in using state-of-the-art CAE hardware and software have crept downward sufficiently so that even small- and medium-sized companies can now develop products by using integrated computer-aided design and manufacturing systems. In large multinational corporations, design teams and technical information are distributed around the world, and computer networks enable products to be developed 24 hours a day. The Boeing 777, regarded as one of the most sophisticated commercial airliners, is a useful case in point because it was the first to be developed through a paperless computer-aided design process. The 777's development began in the early 1990s, and a new computer infrastructure had to be created for the design teams. Conventional paper-and-pencil drafting services were nearly eliminated. Computer-aided design, analysis, and

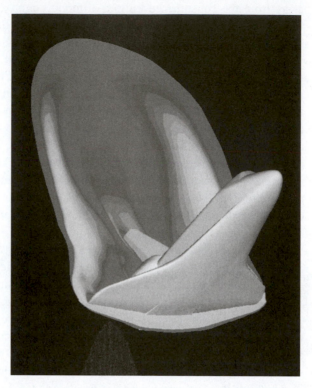

■ **FIGURE 1.20**   Mechanical engineers use computers to analyze airflow around the Space Shuttle orbiter during simulated flight.
Source: Reprinted with permission of NASA.

manufacturing activities were integrated across some 200 design teams that were spread over 17 time zones. Because the plane had over 3 million individual components, making everything fit together was a remarkable challenge, but the CAE tools allowed the designers to check part-to-part fits in a virtual environment before production even began. By constructing and testing fewer physical mock-ups and prototypes, significant development time and cost were saved, and the product was brought to market quicker than would have been possible otherwise (Figures 1.20–1.24).

9. **Bioengineering.** Bioengineering links traditional engineering fields and the life sciences. It applies engineering principles, analysis methods, and design to solve problems occurring in biological systems. Although it is considered an emerging field, bioengineering ranked in the ASME top-ten list not only for the advances that have already been made but also for its future possibilities.

One objective of bioengineering is to create technologies that can support the health care industry, including ultrasound imaging, artificial joint replacement, pacemaker systems, heart valves, minimally invasive surgery techniques, and cryosurgery, in which

■ **FIGURE 1.21**   A scale model of a Vertical Short Takeoff and Landing fighter is mounted in a subsonic wind tunnel. The amount of physical testing that is needed for aircraft design has been significantly reduced through the use of computer-aided engineering analysis.
Source: Reprinted with permission of NASA.

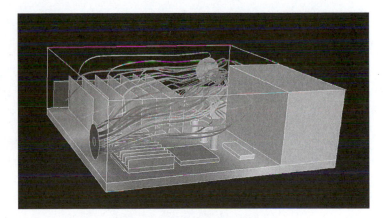

■ **FIGURE 1.22**   Mechanical engineers design the layout and ventilation system for the electronic components in a computer to prevent them from overheating.
Source: Reprinted with permission of Fluent Inc.

the ultralow temperatures produced by liquid nitrogen are used to destroy malignant tumors without damaging the surrounding healthy organs. Tissue engineering is another field where mechanical engineers contribute, and they often work with physicians and scientists to restore damaged skin, bone, and cartilage in the human body (Figures 1.25, 1.26).

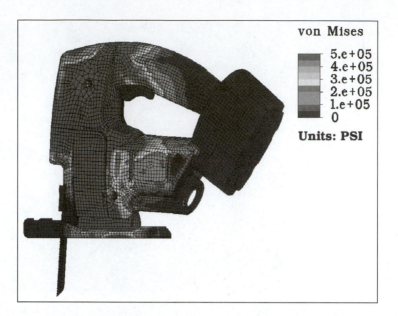

■ **FIGURE 1.23**   Computer simulations enable mechanical engineers to analyze the strength of machines before they are built, as in this analysis of a jigsaw.
Source: Reprinted with permission of Algor Corporation.

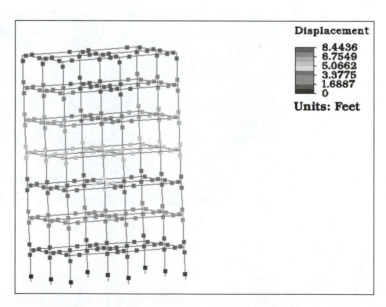

■ **FIGURE 1.24**   The side-to-side vibration of a multistory building during an earthquake is simulated by engineers before construction in order to verify that the building codes will be met.
Source: Reprinted with permission of Algor Corporation.

■ **FIGURE 1.25**   Mechanical engineers design and build automated test equipment that is used in the biotechnology industry. The robotic arm shown here moves a microplate containing samples of deoxyribonucleic acid (DNA) and other chemical compounds during genomic and pharmaceutical discovery research.
Source: Reprinted with permission of Velocity11.

10. **Codes and standards.** Engineers must design components that connect to and are compatible with the hardware developed by others. Because there are codes and standards, you have confidence that a stereo will plug into an electrical outlet in California as well as in Florida and that the voltage will be the same, that the gasoline purchased next month will work in your car just as well as the fuel purchased today, and that the socket wrench purchased at an automobile parts store in the United States will fit the spark plugs on a vehicle manufactured in Germany. Codes and standards are necessary to specify the physical characteristics of mechanical parts so that others can clearly understand their structure and operation.

Many standards are developed through consensus between governments and industry groups, and they have become increasingly important as companies compete internationally for business. Codes and standards involve a collaboration among trade associations, professional engineering societies such as the American Society of Mechanical Engineers, testing groups such as Underwriters Laboratories, and organizations such as the American Society of Testing and Materials. The safety of bicycle and motorcycle helmets, the crash protection features of automobiles and child safety seats, and the fire resistance of home insulation are just some applications for which these guidelines help engineers to design safe products.

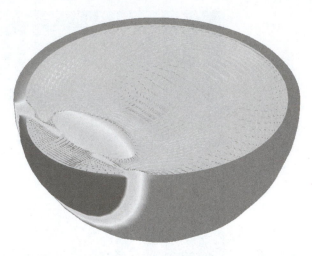

■ **FIGURE 1.26**   Laser iridectomy is a surgical procedure for treating glaucoma in the human eye by equalizing fluid pressure between the eye's anterior and posterior chambers. In collaboration with ophthalmologists, mechanical engineers simulate the temperature of the eye during surgery. The objective of the analysis is to prevent burns on the cornea by better controlling the laser's power and position during the procedure.
Source: Reprinted with permission of Fluent Inc.

1.4   CAREER PATHS

Now that we have introduced the field of mechanical engineering and discussed some of the profession's contributions and technologies, we turn next to the issues of employment options and career growth. Mechanical engineers can work as designers, researchers, and technology managers for companies that range in size from small start-ups to large multinational corporations. Being involved in nearly every stage of a product's life cycle from concept through final production, engineers often work as designers in specifying components, dimensions, materials, and machining processes. A mechanical engineer who specializes in manufacturing is concerned with the production of hardware on a day-to-day basis and with assuring consistent quality and low part-to-part variability. A research and development engineer, on the other hand, works over a longer time frame and is responsible for demonstrating new products and technologies at the proof-of-concept level. Many mechanical engineers work in capacities that are more business than technically oriented. Engineering managers, for instance, organize complex technical operations and help to identify new customers, markets, and products for their company. Figure 1.27 depicts the range of technical and business responsibilities that mechanical engineers can have.

Approximately half of all mechanical engineers work in areas that are related to the manufacturing of machinery, transportation equipment, electrical equipment,

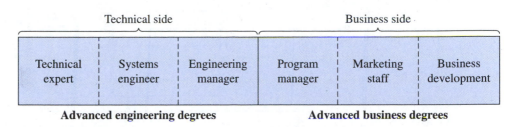

Technical side			Business side		
Technical expert	Systems engineer	Engineering manager	Program manager	Marketing staff	Business development
Advanced engineering degrees			**Advanced business degrees**		

■ **FIGURE 1.27**   A continuum of technical and business responsibilities in engineering.

instruments, and other metal products. A smaller fraction works in public service for such governmental agencies as the National Aeronautics and Space Administration, Department of Defense, National Institute of Standards and Technology, Environmental Protection Agency, and national research laboratories. The remainder work in the fields of consulting, law, education, medicine, and finance. For instance, an investment analyst who evaluates technology-oriented companies might have studied mechanical engineering and would apply that knowledge to better evaluate a company's productivity and performance. Similarly, a patent lawyer who deals frequently with mechanical devices might have a first degree in mechanical engineering. To give you some idea of the range of opportunities available, mechanical engineers can

- Design the motor in a notebook computer's hard disk drive so that it will function reliably for years, even if the computer is accidentally dropped.
- Improve the crash safety of automobiles by analyzing passive seat belt and air bag restraints or by incorporating crumple zones into the vehicle's frame to absorb energy during a collision.
- Upgrade a utility company's electrical power plant in order to improve its efficiency and reduce pollution in the environment.
- Test the apparatus for life science experiments performed on board the Space Shuttle or International Space Station.
- Using aerodynamic principles, determine the shape of turbine blades in a jet engine or select the material alloys from which they will be made.
- Maintain the robotic manufacturing lines that produce such consumer electronics products as computers and personal digital assistants.
- Design the hand and power tools that are used by orthopedic surgeons to perform total hip or knee replacement surgeries.

Career ladder
----------------->

Figure 1.28 illustrates the progression that an individual's career might take after starting work at a hypothetical engineering or technology-oriented company. The terminology and responsibilities for each position will vary from employer to employer, so bear in mind that the descriptions given here are generic. The entry-level position in Figure 1.28 is called the "engineer," and one would begin work in that position after having completed formal study through an accredited

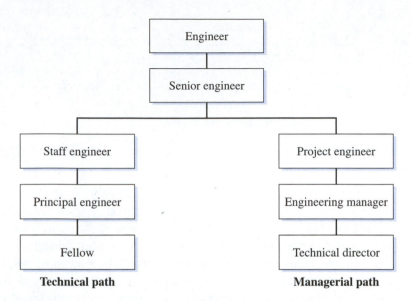

FIGURE 1.28   Potential engineering career opportunities in a hypothetical company.

mechanical engineering program. The engineer works under the supervision of more experienced staff and, in early career stages, must become familiar with the product, customer, design, and manufacturing issues involved. Some of the engineer's responsibilities might include supporting a design team, performing laboratory tests, supervising technicians, and defining performance requirements. After a period of perhaps five years, the engineer's first promotion would be to the level of "senior engineer," a position that entails additional responsibility. A senior engineer could take the leadership role on a design team and be employed in that position for an additional five years or more.

Moving along the path of Figure 1.28, a break point usually occurs in an engineer's career after ten to fifteen years of service and one of two paths will be followed: either a track that emphasizes technology or one that focuses on administration and

Technical track
- - - - - - - - - - - - - ➤

management. Those who follow the managerial path are often interested in business planning and personnel development. Both the technical and managerial paths are essential to a company's success, and one track cannot survive without the other. Mechanical engineers can contribute equally well to both, and the decision as to which track is preferable is an individual decision based on personal

Management track
- - - - - - - - - - - - - ➤

skills and long-term career goals. Generally, the salary scale and recognition within the company are comparable for equivalent levels in the technical and managerial tracks.

The "staff engineer" and "project engineer" are the first parallel levels immediately beyond the break point in Figure 1.28. Potentially, one could serve the company and remain at either of those positions for an indefinite period of time. Staff engineers are responsible for developing the core technologies used in a company's products. The project engineer, on the other hand, manages a group of engineers and is concerned not only with their technical work but also with their professional growth.

Proceeding to the next level in Figure 1.28, the "principal engineer" is regarded as the company's leading technical expert in a certain area, and he or she serves as a resource on which a variety of teams, groups, and divisions can draw. On the parallel business track, the "engineering manager" works at a more senior level than the project engineer and is responsible for allocating and distributing staff and laboratory resources over a range of projects according to the company's business priorities. At the uppermost levels of the parallel technical and management tracks are the ranks of "technical fellow" and "director." The technical fellow is a trailblazer of new technology that will be central to the company several years down the road. The director is responsible for overall administration and brings together engineering and technical staff from multiple disciplines, such as mechanical engineering, electrical engineering, materials engineering, physics, and computer science.

Aside from technical and engineering matters, your progression along the career ladder will depend on a number of skills that, at first glance, might appear to you to be nontechnical in nature. Employers will place high value on your ability to demonstrate initiative when handling work assignments, to work independently and find answers to unexpected problems that develop once a project is underway, to solve problems even in the presence of uncertain or ambiguous data, and to accept additional responsibility

Communication skills with success. Although often underappreciated by beginning students of engineering, communication skills are a major measure of performance in the workplace. Engineering is not a task undertaken exclusively within offices, computer clusters, laboratories, machine shops, and on the factory floor. At each stage of product development, mechanical engineers work with people—supervisors, peers, customers, investors, and vendors. Your abilities to discuss and explain technical and business concepts clearly and effectively and to work with and motivate co-workers, will be critical. The most effective engineers are able to make clear technical presentations, and they document their designs and test results in well-written reports that others read and use. After all, if you have an outstanding and innovative technical idea but you are unable to convey that idea to others in a convincing manner, it is unlikely that the idea will be accepted.

1.5   TYPICAL PROGRAM OF STUDY

As you begin to study mechanical engineering, your college or university program will most likely include (1) general education courses in the areas of the humanities, social sciences, and the fine arts; (2) preparatory courses in mathematics, science, and computer programming; (3) a sequence of core courses in the fundamental subjects that all mechanical engineers should know; and (4) elective courses on specialized topics that you find particularly interesting.

Design and manufacturing ------------▶ The major branches of a generic mechanical engineering curriculum are shown in Figure 1.29. The design and manufacturing branch encompasses the selection of machine components, conceptual design, computer-aided engineering, and the principles of manufacturing. An engineer working in this area might produce the detailed drawings and reports that are necessary to document the design of a new mechanism for feeding sheet paper through

Mechanical systems ------------▶ an ink-jet printer. "Mechanical systems" is the set of subjects encompassing forces, material properties, motion, mechanisms, and mechanical vibration. An engineer working in this field might lead a team that designs an aircraft frame made from lightweight but strong composite materials. In the branch

Thermal-fluid engineering ------------▶ called "thermal-fluid engineering" engineers evaluate efficient methods of energy conversion, power generation, propulsion, and combustion. A thermal-fluids engineer could be involved with improving the performance of an automobile's fuel injection system in order to balance the competing requirements of power production, fuel economy, and air pollution.

Core courses ------------▶ At the next level in Figure 1.29, the exemplary mechanical engineering curriculum branches into more detailed areas of study. This sequence of courses is generally completed in the second through fourth years of study, and it encompasses the following areas:

- Machine and product design—conceiving a new device and designing it in detail.
- Computer-aided engineering—experience with computer-aided design, analysis, and manufacturing tools.

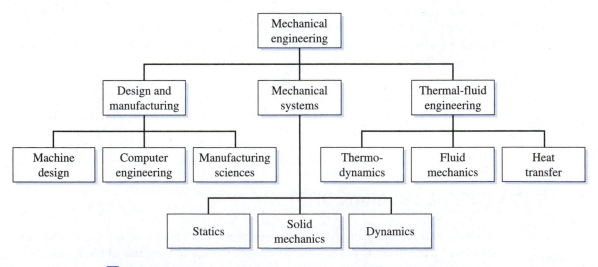

■ **FIGURE 1.29**   Hierarchy of courses completed in a typical mechanical engineering program of study.

- Manufacturing sciences—machining processes, quality control, and production techniques.
- Statics—understanding the forces that act on machines and structures.
- Solid mechanics—stresses and deflections of structural components and choosing the materials from which they are made.
- Dynamics—motion of machines and the forces that are needed to make them move in a desired manner.
- Fluid mechanics—physical properties of different liquids and gases and drag, lift, and buoyancy forces.
- Thermodynamics—energy conservation and conversion and the efficiency of engines and power-generation machinery.
- Heat transfer—conduction, convection, and radiation, heat exchangers, insulation, and cooling systems.

After completing the core curriculum, you will next build an individualized program of study through elective courses. Electives can be chosen to develop specialization within mechanical engineering in such fields as aerospace, automotive, computer-aided design, manufacturing, biomedical engineering, and robotics, among others.

Gaining experience ----------------→ Along with formal education, it is also important for you to gain experience through such activities as summer employment, internships, research projects, and study-abroad opportunities. Those experiences, as well as courses completed outside of the engineering college, will greatly broaden your education and perspective. Increasingly, employers are looking for engineering graduates who have capabilities above and beyond traditional technical skills. Knowledge of business, interpersonal relationships, foreign languages, and communication are important factors for many engineering career choices. For instance, a corporation with overseas subsidiaries, a smaller company that has customers in foreign countries, or a company that purchases equipment from an overseas vendor each value engineers who are conversant in foreign languages. As you begin your study of engineering, be sure to pay attention to those skills.

The Accreditation Board for Engineering and Technology is an organization formed from over two dozen technical and professional societies, including the American Society of Mechanical Engineers. Responsible for accrediting the educational programs that grant engineering degrees in the United States, the board has identified a number of skills that new engineering graduates are expected to have. These criteria are also useful benchmarks for monitoring progress and intellectual growth during your studies.

1. *An ability to apply knowledge of mathematics, science, and engineering.* Since the Second World War, science has been a mainstay of engineering education, and mechanical engineering students in particular have studied mathematics, physics, and chemistry. Students complete core classes that span the classical fields of

thermodynamics, fluid mechanics, heat transfer, solid mechanics, electromechanical systems, dynamics, and controls. The major goal of such courses is to enable you to use basic physical principles for the purpose of understanding, analyzing, and designing practical engineering hardware.

2. *An ability to design and conduct experiments, as well as to analyze and interpret data.* Mechanical engineers must know how to set up and perform experiments, use state-of-the-art instrumentation and measurement equipment, and interpret the physical meaning and implications of the test results.

3. *An ability to design a system, component, or process to meet desired needs.* This skill lies at the very heart of mechanical engineering. Your objectives are to learn systematic procedures for conceptualizing solutions to open-ended problems, for recognizing the level of analysis required to select components and prepare a detailed design, and for producing functional hardware.

4. *An ability to function on multidisciplinary teams.* Mechanical engineering is not a solo activity, and you will need to develop the skills necessary to interact effectively with others in the business community. In industry, products and hardware are generally developed by technical teams, not by individuals. Further, the team may comprise engineers from various disciplines, scientists, marketing staff, and others. At the most fundamental level, the ability to be a team player requires basic interpersonal and communication skills. One important aspect of communication is a shared technical vocabulary, and, after completing this textbook, you will be able to discuss many technical topics of interest to mechanical engineers.

5. *An ability to identify, formulate, and solve engineering problems.* Engineering is firmly based on mathematical and scientific principles, but some art is also involved as those methods are applied to build something new. Engineers are known as "problem solvers," a reputation that requires such abilities as confidence when confronting unfamiliar situations, defining the problem clearly, making rough estimates, dealing with the reality of multiple solutions, and communicating results to others.

6. *An understanding of professional and ethical responsibility.* Through your courses and personal experiences, you will see that engineers have a responsibility to act professionally and ethically. Engineers must be able to recognize ethical and business conflicts and to resolve them when they arise.

7. *An ability to communicate effectively.* Engineers are expected to be competent in both written and verbal communication, including the presentation of engineering calculations, computations, measurement results, and designs.

8. *The broad education necessary to understand the impact of engineering solutions in a global and societal context.* Engineering is not practiced in a vacuum, and an engineer's work product will be used by others, perhaps on a wide scale. A glance at the top-ten list of Section 1.3 reinforces the perspective that engineers should view their efforts as intended to advance society. An engineer who is aware of and knowledgeable about that context will be more effective in making sound engineering, ethical, and career decisions.

9. *A recognition of the need for and an ability to engage in lifelong learning.* Your intellectual growth will continue long after graduation and, to stay up-to-date with evolving technology, should extend throughout your entire career.

10. *A knowledge of contemporary issues.* Engineers need to be aware of developments that are of current importance to society at large as well as to the engineering profession.

11. *An ability to use the techniques, skills, and modern engineering tools necessary for engineering practice.* This skill is based in part on using computer-aided engineering analyses, the ability to think critically about numerical results, and the ability to program in different computer languages.

It is worthwhile to keep this skill set in mind as you begin your formal study of engineering and to reflect periodically on how the material in your courses contributes to your professional development.

SUMMARY

Engineers conceive, design, analyze, and build things that work. Engineers are regarded as being efficient, logical, and organized problem solvers. Mechanical engineering is a diverse field, and it is the most general of the traditional engineering disciplines. The products and systems produced by mechanical engineers have improved the day-to-day lives of billions of people, in a very literal sense. To accomplish their work, mechanical engineers use computer-aided tools for conceptual design, detailed component design, performance simulation, and manufacturing. Technologies that you may have previously taken for granted—such as abundant and inexpensive electricity, refrigeration, and transportation—take on new meaning as you reflect on their importance to our society and on the remarkable hardware that makes them possible.

SELF-STUDY AND REVIEW

1. Define the term *engineering*.
2. Discuss the differences between engineers and mathematicians and scientists.
3. Define the term *mechanical engineering*, and discuss it relative to the other traditional engineering disciplines.
4. Describe a half-dozen products that mechanical engineers design, improve, or produce, and list some technical issues that must be solved.
5. Describe several significant achievements of the mechanical engineering profession.
6. Discuss the technical and management career paths that are available to mechanical engineers.
7. Describe the main areas that comprise a typical mechanical engineering curriculum.

PROBLEMS

1. For each of the following systems, give two examples of how a mechanical engineer would be involved in design, manufacture, analysis, or testing. For each of your two examples, write several sentences or make a drawing to explain the contribution.

   - Passenger automobile engine
   - League-approved baseball batting helmet
   - Computer hard disk drive
   - Artificial hip implant
   - Microprocessor chip

2. Repeat the exercise of Problem 1 for the following systems:

   - Jet engine for a commercial airliner
   - Rover robot for planetary exploration
   - Computer ink-jet printer
   - Mobile telephone
   - Can of soda from a vending machine

3. Repeat the exercise of Problem 1 for the following systems:

   - Graphite epoxy skis, tennis racket, or golf club
   - Elevator or escalator
   - Compact disc player
   - Automatic teller banking machine
   - Automotive child safety seat

4. Repeat the exercise of Problem 1 for the following systems:

   - Hybrid gas-electric passenger vehicle
   - Motors for remote control cars, planes, and boats
   - Bindings for snowboards
   - Global Positioning System (GPS) satellite receiver
   - Motorized wheelchair

5. Repeat the exercise of Problem 1 for the following systems:

   - Orbiting communications satellite
   - Microwave oven
   - Aluminum mountain bike frame
   - Cordless electric drill
   - Automotive antilock brake system

6. Read one of the articles, listed in the References section at the end of this chapter, which appeared in *Mechanical Engineering* magazine and describes a top-ten achievement. Prepare an approximately 250-word report on the article summarizing the interesting and important aspects of the achievement. The write-up should be prepared on a word processor.

7. What other product or device do you think should be on the list of the top mechanical engineering achievements? Prepare an approximately 250-word report detailing your rationale and listing the interesting and important aspects of the achievement. The write-up should be prepared on a word processor.

8. Looking forward by 100 years, what technological advance do you think will be regarded at that time as having been a significant achievement of the mechanical engineering profession during the twenty-first century? Prepare an approximately 250-word report that explains the rationale for your speculation. The write-up should be prepared on a word processor.

9. What do you think are the three most significant issues facing engineers today? Prepare an approximately 250-word report that explains your rationale. The write-up should be prepared on a word processor.

10. Interview someone whom you know or contact a company and learn some of the details behind a product that interests you. Prepare an approximately 250-word summary of the product, the company, or the manner in

which a mechanical engineer contributes to the product's design or production. The write-up should be prepared on a word processor.

11. Interview someone whom you know or contact a company and learn about a computer-aided engineering software tool that is related to mechanical engineering. Prepare an approximately 250-word summary that describes what the software tool does and how it can help an engineer work more efficiently and accurately. The write-up should be prepared on a word processor.

12. Find a newspaper or magazine article pertaining to a technical or engineering subject that has been in the news recently. Prepare an approximately 250-word report on the issue. Describe how mechanical engineers are or could be involved with investigating the issue or solving the problem at hand. The write-up should be prepared on a word processor.

REFERENCES

Armstrong, N. A., "The Engineered Century," *The Bridge*, National Academy of Engineering, Spring 2000, pp. 14–18.

Gaylo, B. J., "That One Small Step," *Mechanical Engineering*, ASME International, October, 2000, pp. 62–69.

Ladd, C. M., "Power to the People," *Mechanical Engineering*, ASME International, September, 2000, pp. 68–75.

Lee, J. L., "The Mechanics of Flight," *Mechanical Engineering*, ASME International, July, 2000, pp. 55–59.

Leight, W. and Collins, B., "Setting the Standards," *Mechanical Engineering*, ASME International, February, 2000, pp. 46–53.

Lentinello, R. A., "Motoring Madness," *Mechanical Engineering*, ASME International, November, 2000, pp. 86–92.

Nagengast, B., "It's a Cool Story," *Mechanical Engineering*, ASME International, May, 2000, pp. 56–63.

Petroski, H., "The Boeing 777," *American Scientist*, November–December, 1995, pp. 519–522.

Rastegar, S., "Life Force," *Mechanical Engineering*, ASME International, March, 2000, pp. 75–79.

Rostky, G., "The IC's Surprising Birth," *Mechanical Engineering*, ASME International, June, 2000, pp. 68–73.

Schueller, J. K., "In the Service of Abundance," *Mechanical Engineering*, ASME International, August, 2000, pp. 58–65.

United States Department of Labor, Bureau of Labor Statistics, *Occupational Outlook Handbook*, 2000–2001 Edition (Bulletin 2520).

Weisberg, D. E., "The Electronic Push," *Mechanical Engineering*, ASME International, April, 2000, pp. 52–59.

2 Problem-Solving Skills

2.1 OVERVIEW

In this chapter we begin discussing some of the steps that engineers follow when they solve problems and perform calculations in their daily work. Mechanical engineers are fluent with numbers, and they are organized when obtaining numerical answers to questions that involve a remarkable breadth of variables and physical properties. Some of the quantities that you will encounter in your study of mechanical engineering are force, torque, thermal conductivity, shear stress, fluid viscosity, elastic modulus, kinetic energy, Reynolds number, specific heat, and so on. The list is very long indeed. The only way that you will make sense of so many quantities is to be very clear about them in calculations and when explaining results to others.

Each quantity in mechanical engineering has two components: a numerical value and a unit. One is simply meaningless without the other, and practicing engineers pay as close and careful attention to the units in a calculation as they do to the numbers. In the first portion of this chapter, we will discuss fundamental concepts for systems of units and conversions between them and a procedure for checking dimensional consistency that will serve you well in engineering calculations.

Aside from the question of units, engineers must obtain numerical answers to questions—how strong? how heavy? how much power? what temperature?—often in the face of uncertainty and incomplete information. At the start of a new design, for instance, the shape and dimensions of the final product are not known; if they were, then it wouldn't be necessary to design in the first place. Only rough estimates of the forces applied to a structure might be available. Likewise, exact values for material properties are rarely known, and there will always be some variation between samples of materials. Nevertheless, an engineer still has a job to do, and the design process must start somewhere.

For that reason, engineers make approximations in order to assign numerical values to quantities that are otherwise unknown. Those approximations are understood to be imperfect, but they are better than random guesses and certainly better than nothing. Mechanical engineers use their common sense, experience, intuition, judgment, and physical laws to find answers through a process called *order-of-magnitude approximation*. These estimates are a first step toward reducing a concrete physical situation to its most essential and relevant pieces. Mechanical engineers are comfortable making reasoned approximations.

In the following sections we begin a discussion of numerical values, unit systems, significant digits, and approximation. These principles and techniques will be applied further as we explore the "elements" of mechanical engineering outlined in later chapters. After completing this chapter, you should be able to:

- Report both a numerical value and its unit in each calculation that you perform.
- List the base units in the United States Customary System and the *Système International d'Unités*, and state some of the derived units used in mechanical engineering.
- Understand the need for proper bookkeeping of units when making engineering calculations and the implications of not doing so.
- Convert numerical quantities from the United States Customary System to the *Système International d'Unités*, and vice versa.
- Check your equations and calculations to verify that they are dimensionally consistent.
- Understand how to perform order-of-magnitude approximations.

2.2   UNIT SYSTEMS AND CONVERSIONS

Engineers specify physical quantities in two different, but conventional, systems of units: the United States Customary System (USCS) and the International System of Units (*Système International d'Unités*, or SI). Practicing mechanical engineers must be conversant with both unit systems. They need to convert quantities from one system to the other, and they must be able to perform engineering calculations equally well in either system. In this textbook examples and problems will be formulated in both systems so that you will be able to develop familiarity and practice with the USCS and SI. As we introduce new quantities and variables in the following chapters, the corresponding USCS and SI units will be described, along with their conversion factors.

You cannot escape conversion between the USCS and SI, and you will not find your way through the maze of mechanical engineering without being proficient with both sets of units. The decision as to whether the USCS or SI should be used when solving a particular problem will depend on how the information was initially specified. If the information is given in the USCS, then you should solve the problem and apply formulas using the USCS alone. Conversely, if the information is given in the SI, then formulas should be applied using the SI alone. It is bad practice for you to do otherwise—namely, to take data given in the USCS, convert to SI, perform calculations in SI, and then convert back to USCS (or vice versa). The reason for this recommendation is twofold. First, from the practical day-to-day matter of being a competent engineer, you will need to be fluent in both the USCS and SI. Second, the additional steps that are involved when quantities are converted from one system to another, and back again, are simply further opportunities for errors to enter into your solution. By way of motivation, the following case study highlights the problems that can arise when unit systems are mixed and an otherwise straightforward calculation is made in error.

Keeping Track of Units on Flight 143

In July of 1983, Air Canada Flight 143 was en route from Montreal to Edmonton. The Boeing 767 had three fuel tanks, one in each wing and one in the fuselage, which supplied the plane's two jet engines. Flying through clear sky on a summer day, remarkably, one fuel pump after another stopped as each tank on the jetliner ran completely dry. The engine on the left wing was the first to stop, and 3 minutes later, as the plane descended, the second engine stopped. Except for small auxiliary backup systems, this sophisticated aircraft was without power.

   The flight crew and air traffic controllers decided to make an emergency landing at an airfield that was at the time abandoned but that was once used by the Royal Canadian Air Force. Through their training and skill, the flight crew was fortunately able to safely land the plane, narrowly missing race cars and spectators on the runway who had gathered that day for amateur auto racing (Figure 2.1). Although the racing

■ **FIGURE 2.1**   Flight 143 following its emergency landing and evacuation on the runway of an abandoned airport, which was being used at the time for automobile races.
Source: Reprinted with permission of the Winnipeg Free Press.

enthusiasts didn't notice the unpowered, quietly approaching, gliding airplane until the very last minute, they were able to get out of the way. Despite the collapse of the front landing gear, and the subsequent damage that occurred to the plane's nose, the flight crew and passengers suffered no serious injuries.

As is the case in incidents involving people and technology, several related causes—some technical and some interpersonal—contributed to the accident. After a thorough investigation, a review board determined that one of the important factors behind the accident was a refueling error in which the quantity of fuel that should have been added to the tanks was incorrectly calculated.

During the flight's preparations, it was determined that 7682 liters (L) of fuel were already in the plane's tanks. To complete the refueling operation, it was necessary to calculate the number of liters that had to be added so that the tanks contained a combined total of 22,300 kilograms (kg) of fuel, the amount known to be needed for the Montreal-to-Edmonton flight. In prior years, the airline had expressed the amount of fuel needed for each flight in the units of pounds (lb); however, fuel consumption of the new 767s was calculated in kilograms. A situation arose in which fuel was measured by volume (L), weight (lb), and mass (kg) in two different systems of units.

In the refueling calculations, the conversion factor of 1.77 had been used when converting the volume of fuel (L) to mass (kg). However, the units associated with that number were not explicitly stated or checked. In fact, the units for the conversion factor stipulate that the density of jet fuel is 1.77 lb/L, not 1.77 kg/L. As a result of the miscalculation, roughly 9000 instead of 16,000 liters of fuel were added to the plane. As Flight 143 took off for western Canada, it was well short of the fuel required for the trip. This case study illustrates several points that you should note, including the issues of accurate unit conversion and designing engineering systems so that they can be operated as simply as possible.

Base and Derived Units

Base units
- - - - - - - - - - - - - - ->
Given some perspective from the story of Flight 143 as to the importance of units and their bookkeeping, we now turn to the specifics of the USCS and SI.

A unit is an arbitrary division of a physical quantity, and it has a magnitude that is agreed upon by mutual consent. Both the USCS and SI are made up of base units and derived units. A base unit is a fundamental quantity that cannot be further broken down or expressed in terms of any simpler elements. Base units are the core building blocks of any unit system. As an example, the base unit for length in the SI is the meter (m), whereas in the USCS, the base unit is the foot (ft). Derived units, as their

Derived units
- - - - - - - - - - - - ->
name implies, are constructed as combinations of base units. An example of a derived unit in the SI is velocity (m/s), which is a combination of the base units for length and time. The mile, defined as 5280 ft, is a derived unit for length in the USCS.

United States Customary System

The United States Customary System of units is a historical and traditional system, and its origin traces back to the ancient Roman Empire. In that vein, the abbreviation for pound (lb) is taken from the Roman unit of weight, *libra*. The USCS includes such measures as pounds, ounces, tons, feet, inches, miles, seconds, gallons, and quarts. The USCS evolved from a unit system originally used in Great Britain, but today it is used primarily in the United States. Most other industrialized countries have adopted the SI as their uniform standard of measurement for business and

USCS

- - - - - - - - - - - - - - ▶ commerce. However, you should note that engineers practicing in the United States, or in companies or governmental agencies having U.S. affiliations, need to be skillful with both the USCS and SI.

Why does the United States stand out in retaining the USCS? The reasons are both logistical and cultural: There is already a vast continent-sized infrastructure within the United States that is based on the USCS. Conversion away from the existing system would be a significant and expensive burden. The dimensions of countless existing structures, factories, machines, and spare parts have already been specified and built in terms of the USCS. Furthermore, while most American consumers are comfortable purchasing, say, a gallon of gasoline and most drivers have an intuitive feel for how fast 50 mph seems in an automobile, they are not as familiar with the SI counterparts. That being noted, standardization to the SI in the United States is proceeding, ever so slowly, because of the need for companies to interact with and compete against their counterparts in the international business community. Until such time as the United States has made a full transition to the SI (and don't hold your breath), it will be necessary—and indeed essential—for you to be proficient with both unit systems.

Tables 2.1 and 2.2 list the base units and certain common derived units in the USCS. There are only seven base units: foot, pound, second, ampere, Rankine, mole, and candela. On the other hand, many derived units are formed as combinations of these base units. In particular, and as a common point of potential error, you should

TABLE 2.1   Base Units in the USCS

Quantity	USCS Base Unit	Abbreviation
Length	foot	ft
Force	pound	lb
Time	second	s
Electric current	ampere	A
Thermodynamic temperature	Rankine	R
Amount of substance	mole	mol
Luminous intensity	candela	cd

TABLE 2.2   Certain Derived Units in the USCS. Although a change in temperature of 1 Rankine also equals a change of 1 degree Fahrenheit, the absolute values are converted using the formula.

Quantity	USCS Derived Unit	Abbreviation	Definition
Length	mil	mil	1 mil = 0.001 in.
Length	inch	in.	1 in. = 0.0833 ft
Length	yard	yd	1 yd = 3 ft
Length	mile	mi	1 mi = 5280 ft
Volume	U.S. gallon	gal	1 gal = 0.1337 ft^3
Mass	slug	slug	1 slug = 1 lb·s^2/ft
Force	ounce	oz	1 oz = 0.0625 lb
Force	ton	ton	1 ton = 2000 lb
Moment of a force	ft·lb	ft·lb	—
Pressure or stress	psi	psi	1 psi = 1 lb/in^2
Energy, work, or heat	ft·lb	ft·lb	—
Energy, work, or heat	Btu	Btu	1 Btu = 778.2 ft·lb
Power	horsepower	hp	1 hp = 550 ft·lb/s
Temperature	degree Fahrenheit	°F	°F = °R − 460

note that the pound (the unit of force) is a base unit in the USCS. The unit of mass in the USCS (which is called the slug) is actually a derived unit that is equivalent to 1 lb·s^2/ft. We will discuss calculations in the USCS and SI involving mass and force through the following examples.

International System of Units

The International System of units (*Système International d'Unités*, or SI) is the measurement standard based in part on the quantities of meters, kilograms, and seconds.

SI - - - - - - - - - - - - - - - → Tables 2.3 and 2.4 summarize its base units and certain derived ones. Conversion factors between the USCS and the SI are listed in Tables 2.5 and 2.6.

The base units in the SI are set using standards that are defined by international agreements. The origins of the meter trace back to the eighteenth century; it was originally intended to be equivalent to 1 ten-millionth (10^{-7}) of the length of the meridian passing through Paris from the pole to the equator (one-quarter of the Earth's circumference). Later, the standard meter was defined by the length of a bar made from a platinum–iridium metal alloy. Copies of the bar, called *prototypes*, were distributed to laboratories around the world. By mutual international agreement, the length of the bar was always measured at the temperature of melting water ice.

TABLE 2.3   Base Units in the SI

Quantity	SI Base Unit	Abbreviation
Length	meter	m
Mass	kilogram	kg
Time	second	s
Electric current	ampere	A
Thermodynamic temperature	Kelvin	K
Amount of substance	mole	mol
Luminous intensity	candela	cd

TABLE 2.4   Certain Derived Units in the SI. Although a change in temperature of 1 Kelvin equals a change of 1 degree Celsius, the absolute values are converted using the formula.

Quantity	SI Derived Unit	Abbreviation	Definition
Length	micron	μm	$1\ \mu m = 10^{-6}$ m
Volume	liter	L	$1\ L = 0.001$ m^3
Force	newton	N	$1\ N = 1$ kg·m/s^2
Moment of a force	N·m	N·m	—
Pressure or stress	pascal	Pa	$1\ Pa = 1$ N/m^2
Energy, work, or heat	joule	J	$1\ J = 1$ N·m
Power	watt	W	$1\ W = 1$ J/s
Temperature	degree Celsius	°C	°C = °K − 273

TABLE 2.5   Conversion Factors Between Certain Quantities in the USCS and SI

Quantity	To Convert from ...	To ...	Multiply by ...
Length	foot (ft)	meter (m)	0.3048
Length	inch (in.)	meter (m)	0.0254
Length	mile (mi)	kilometer (km)	1.609
Volume	gallon (gal)	meter^3 (m^3)	3.785×10^{-3}
Volume	gallon (gal)	liter (L)	3.785

TABLE 2.5	*(continued)*		

Quantity	To Convert from ...	To ...	Multiply by ...
Mass	slug	kilogram (kg)	14.59
Force	pound (lb)	newton (N)	4.448
Pressure or stress	pound/inch^2 (psi)	pascal (Pa)	6895
Work, energy, or heat	foot-pound (ft·lb)	joule (J)	1.356
Power	foot-pound/second (ft·lb/s)	watt (W)	1.356
Power	horsepower (hp)	kilowatt (kW)	0.7457

TABLE 2.6	Conversion Factors Between Certain Quantities in the SI and USCS		

Quantity	To Convert from ...	To ...	Multiply by ...
Length	meter (m)	foot (ft)	3.281
Length	meter (m)	inch (in.)	39.37
Length	kilometer (km)	mile (mi)	0.6214
Volume	meter^3 (m^3)	gallon (gal)	264.2
Volume	liter (L)	gallon (gal)	0.2642
Mass	kilogram (kg)	slug	0.0685
Force	newton (N)	pound	0.2248
Pressure or stress	pascal (Pa)	pound/inch^2 (psi)	1.450×10^{-4}
Work, energy, or heat	joule (J)	foot-pound (ft·lb)	0.7376
Power	watt (W)	foot-pound/second (ft·lb/s)	0.7376
Power	kilowatt (kW)	horsepower (hp)	1.341

The definition of the meter has been periodically updated in order to make the length standard more robust and repeatable, all the while changing the actual length by as little as possible. As of October 20, 1983, the meter was established as the length of the path traveled by light in vacuum during a time interval of 1/299,792,458 of a second, which is measured to high accuracy by an atomic clock.

At the end of the eighteenth century, the kilogram was defined as the mass of 1000 cm^3 of water. Today, the kilogram is determined by the mass of a physical sample that is called the *standard kilogram* and is also made of platinum and iridium. Thus, while the meter is based on a reproducible measurement involving the speed of light and time, the kilogram is still defined by an actual physical sample.

TABLE 2.7 Order-of-Magnitude Prefixes in the SI		
Name	**Symbol**	**Multiplicative Factor**
tera	T	$1,000,000,000,000 = 10^{12}$
giga	G	$1,000,000,000 = 10^{9}$
mega	M	$1,000,000 = 10^{6}$
kilo	k	$1000 = 10^{3}$
hecto	h	$100 = 10^{2}$
deca	da	$10 = 10^{1}$
deci	d	$0.1 = 10^{-1}$
centi	c	$0.01 = 10^{-2}$
milli	m	$0.001 = 10^{-3}$
micro	μ	$0.000,001 = 10^{-6}$
nano	n	$0.000,000,001 = 10^{-9}$
pico	p	$0.000,000,000,001 = 10^{-12}$

SI units are commonly combined with a prefix so that the numerical value that is written does not have a power-of-10 exponent that is either too large or too small. You should use a prefix to shorten the representation of numerical values and to condense an otherwise excessive number of trailing zero digits in your calculations.

Prefix ----------------→ The standard prefixes in the SI are listed in Table 2.7. It is good practice not to use a prefix for any numerical value that falls between 0.1 and 1000. On the other hand, when a numerical quantity is either very small or very large, a prefix will represent it in a clean, compact, and conventional manner.

EXAMPLE 2.1

The KC-10 "Extender" aerial tanker aircraft of the United States Air Force is used to refuel other planes in flight. The KC-10 can carry 365,000 lb of jet fuel, which is transferred to another aircraft through a boom that temporarily connects the two planes. (a) Express the mass of the fuel in the USCS. (b) Express the mass of the fuel in the SI. (c) Express the weight of the fuel in the SI.

SOLUTION

(a) We note that 365,000 lb is the weight of the fuel, and we convert from weight W to mass m through the definition

$$W = mg$$

where g is the gravitational acceleration constant ($g = 32.2$ ft/s^2 in the USCS and

$g = 9.81$ m/s^2 in the SI). The mass of the fuel is therefore

$$m = \frac{W}{g} = \frac{3.65 \times 10^5 \text{ lb}}{32.2 \text{ ft/s}^2} = 1.134 \times 10^4 \text{ lb·s}^2/\text{ft} = 1.134 \times 10^4 \text{ slugs}$$

(b) We are asked to convert the mass quantity 1.134×10^4 slugs into the corresponding units of kg in the SI. No conversion factor exists directly between lb (as given) and kg (as desired). Those units correspond to force and mass, respectively, which are two different physical quantities. Referring to Table 2.5, we see that

1 slug $= 14.59$ kg

and the fuel's mass is calculated to be

$$m = (1.134 \times 10^4 \text{ slugs})(14.59 \text{ kg/slug}) = 1.655 \times 10^5 \text{ kg}$$

(c) The fuel's weight in the SI is

$$W = (1.655 \times 10^5 \text{ kg})(9.81 \text{ m/s}^2) = 1.62 \times 10^6 \text{ N} = 1.62 \text{ MN}$$

where we have used the SI prefix M (mega-) to represent the factor of 1 million. To double-check these calculations, we can convert the fuel's 365,000 lb weight directly to the SI. By using Table 2.5, we see that

1 lb $= 4.448$ N

so that in the SI,

$$W = (3.65 \times 10^5 \text{ lb})(4.448 \text{ N/lb}) = 1.62 \times 10^6 \text{ N} = 1.62 \text{ MN}$$

confirming our previous number.   ∎

EXAMPLE 2.2

Helium–neon (or He–Ne) lasers are used in engineering laboratories, robot vision systems, and even in the bar code readers within checkout counters at supermarkets. A certain He–Ne laser has a power output of 3 mW and produces light with a wavelength of 632.8 nm. (a) Convert the power rating to horsepower. (b) Convert the wavelength to inches.

SOLUTION

(a) Referring to Table 2.7, the notation mW refers to a milliwatt, or 10^{-3} W. Thus, the laser produces 3×10^{-3} W $= 3 \times 10^{-6}$ kW. Using the conversion factor between kW and hp that is listed in Table 2.6, we have

$$P = (3 \times 10^{-6} \text{ kW})(1.341 \text{ hp/kW}) = 4.02 \times 10^{-6} \text{ hp}$$

The unit of horsepower is not necessarily convenient for describing the laser's output because that unit is so much larger than the power level.

(b) Referring to Table 2.7, one nanometer (nm) is equivalent to one-billionth of a meter. The conversion for the laser's wavelength becomes

$$\lambda = (632.8 \times 10^{-9} \text{ m})(39.37 \text{ in./m}) = 2.49 \times 10^{-5} \text{ in.}$$

The Greek letter lambda (λ) is a conventional mathematical symbol used for wavelength. Appendix A summarizes the names and symbols of Greek letters.   ∎

2.3 DIMENSIONAL CONSISTENCY

When you apply equations of mathematics, science, or engineering, the calculations must be dimensionally consistent, or they are wrong. "Dimensional consistency" means that the units associated with the numerical values on each side of the equality sign match. Likewise, if two terms are combined in an equation by summation, or they are subtracted from one another, the numerical values must have the same units. This principle is a straightforward means to double-check your algebraic and numerical work.

In paper-and-pencil calculations, it is good practice to keep the units adjacent to each numerical quantity in an equation. Likewise, combine the units of terms, or cancel them, at each step in the solution. You should retain units in the equation and manipulate them just as you would any other algebraic quantity. In that manner, and by using the principle of dimensional consistency, you can use the properties of units to double-check your calculation and develop greater confidence in its accuracy. Of course, it is possible that the result might be incorrect for a reason other than units. Nevertheless, performing a double-check on the units in any equation is always a good idea. The process of verifying the dimensional consistency of an equation by keeping track of units is illustrated through the following example.

EXAMPLE 2.3

In Figure 2.2, the steel bit is held in the chuck of the drill press. The drill bit has diameter $d = 6$ mm (the curved flutes are small and they have been neglected) and length $L = 65$ mm. The bit is accidentally bent as the work piece shifts during a drilling operation, and it is subjected to the side force of $F = 50$ N. As derived in mechanical engineering courses on stress analysis, the sideways deflection of the tip is calculated through the equation

$$\Delta x = \frac{64FL^3}{3\pi E d^4}$$

where each term in the equation has the following units:

Δx (length) is the deflection of the tip.
F (force) is the magnitude of the force applied at the tip.

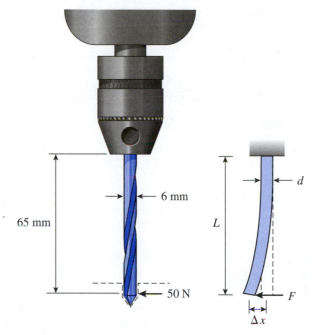

FIGURE 2.2   Sideways bending of a drill bit in Example 2.3.

L (length) is the drill bit's length.
E (force/length^2) is a property of the drill bit's material called the *elastic modulus*.
d (length) is the drill bit's diameter.

(a) Verify that this equation is dimensionally correct. (b) By using the value $E = 200 \times 10^9$ Pa, calculate the amount Δx that the bit's tip deflects.

SOLUTION

(a) The quantity $64/3\pi$ is dimensionless—a pure number—and therefore it has no units to influence the equation's dimensional consistency. For the other terms, the units of each quantity are combined in the equation according to

$$(\text{length}) = \frac{(\text{force})(\text{length})^3}{(\text{force}/(\text{length})^2)(\text{length})^4}$$

$$= \frac{(\text{length})^3}{(\text{length})^2}$$

$$= (\text{length})$$

Because the units on each side of the equation are identical, the equation is indeed dimensionally consistent.

(b) The tip moves sideways by the amount

$$\Delta x = \frac{64FL^3}{3\pi Ed^4}$$

$$= \frac{64(50 \text{ N})(0.065 \text{ m})^3}{3\pi(200 \times 10^9 \text{ Pa})(6 \times 10^{-3} \text{ m})^4}$$

In the next step, we combine numerical values and group units

$$\Delta x = 360 \times 10^{-6} \frac{\text{N·m}^3}{(\text{N/m}^2)(\text{m}^4)}$$

in which we have expanded Pa as N/m^2 in order to cancel units in the denominator. Combining the units algebraically leads to

$$\Delta x = 360 \times 10^{-6} \text{ m}$$

Because that result has a large negative exponent, we convert it to standard form by using the SI prefix μ (or micro) to represent one-millionth:

$$\Delta x = 360 \ \mu\text{m}$$

The tip will move by just over one-third of a millimeter.   ■

Mechanical engineers often work with dimensionless numbers. These quantities are either pure numbers that have no units, or they are groupings of variables in which the units precisely cancel one another, again leaving a pure number. One example with which you might already be familiar is the so-called Mach number Ma which is used to measure the speed of aircraft. It is named after the nineteenth-century physicist Ernst Mach. The Mach number is defined by the equation $Ma = v/c$ and is simply the ratio of the aircraft's speed v to the speed of sound c in air (approximately 700 mph). When the numerical values for both v and c are expressed in the same units (for instance, mph), the units will cancel in the equation for Ma. A commercial airliner might cruise at a speed of $Ma = 0.7$, while a supersonic fighter could travel at $Ma = 1.4$. As a cautionary note, it would be incorrect in the expression for Ma to specify v and c with different units, say, m/s and km/h, respectively. Although either unit would indeed be correct for speed, they would not cancel one another in the calculation of Ma.

*Dimensionless
number*
- - - - - - - - - - - - - - - ->

EXAMPLE 2.4

In such engineering applications as designing the shape of the body of a car or the wing of an aricraft, it is necessary to know the drag force D that will resist high-speed motion through the air. The coefficient of drag, C_D, measures the amount of resistance that an object experiences as it moves through air, or as air flows around it. The drag coefficient is defined by the equation

$$C_D = \frac{D}{\frac{1}{2}\rho A v^2}$$

where ρ is the density of air, A is called the frontal area of the object, and v is its speed. The Greek letter rho (ρ) is a conventional mathematical symbol used for density. Appendix A summarizes the names and symbols of Greek letters. Verify that C_D is a dimensionless number.

SOLUTION

In order that C_D be dimensionless, the units of the denominator must combine to have the unit of force. Thus, examining the term $\rho A v^2$, we have

$$(\text{mass/length}^3)(\text{length}^2)(\text{length/time})^2 = (\text{mass})(\text{length})/(\text{time}^2) = \text{force}$$

confirming that C_D is dimensionless. ■

2.4 SIGNIFICANT DIGITS

A significant digit is a numerical value that is known to be correct and reliable in the light of inaccuracy that is present in the supplied information, any approximations that have been made along the way, and the mechanics of the calculation itself. As a general rule, the last significant digit that you report in the answer to a problem should be of the same order of magnitude as the last significant digit that was supplied in the problem's statement. It would be inappropriate to report more significant digits in the answer than were given in the supplied data, since that implies that the output of a calculation is somehow more accurate than the input to it.

Precision
- - - - - - - - - - - - - - - - →

The precision of a number is half as large as the last significant digit used in expressing the number. The factor of one-half arises because the last digit of a number represents the rounding-off process, either higher or lower, of the trailing digits. For instance, if an engineer records in a design notebook that the force acting on the bearing of a hard disk drive's motor is 43.01 mN, the statement means that the force is closer to 43.01 mN than it is to either 43.00 mN or 43.02 mN. The precision of this numerical value is ±0.005 mN, the difference that could be present in the force reading and still result in a rounded value of 43.01 mN. Even when we write 43.00 mN, a numerical value that has two trailing zeros, four significant digits are still present, and the implied precision remains ±0.005 mN. Alternatively, if the engineer had written the force as 43.010 mN, that would imply that the value is known quite accurately indeed and that the force lies closer to 43.010 mN than to either 43.009 mN or 43.011 mN. The precision is now ±0.0005 mN.

Likewise, you can see how some ambiguity is present when a numerical value is stated as 200 lb. On the one hand, the value could mean that the measurement was made only within the nearest 100 pounds and that the actual force is closer to 200 lb than to either 100 lb or 300 lb. On the other hand, the value could mean that the force is indeed 200 lb, and not 199 lb or 201 lb.

As a general rule, during the intermediate steps of a calculation, you should retain several more significant digits than you expect to eventually report in the final answer. In that manner, rounding errors will not creep into your solution, compound along the way, and distort your final answer. When the calculation is complete, you can then always truncate the answer to a reasonable number of significant digits. These considerations lead us to the following rule-of-thumb guideline:

> For the purpose of examples and problems in this textbook, treat the supplied data as being exact. Recognizing engineering approximations and limits on measurements, though, report your answers to three or four significant digits.

You do need to be cautious of the misleading sense of accuracy that is afforded by the use of calculators and computers. While a calculation can certainly be performed to eight or more significant digits, nearly all dimensions, material properties, and other physical parameters that are encountered in mechanical engineering are known to far fewer digits. Although the computation itself might be very accurate, the input data that would be supplied to the calculation will rarely have the same level of accuracy.

For instance, in analyzing the crash restraint system in an automobile, an important design parameter is the driver's or passenger's weight. An engineer needs to make calculations based on the weight of an average person in an automobile, but clearly, there can be wide variation. Likewise, the numerical value of the friction coefficient between a tire and the road will change depending on the condition of the tires, the presence of dirt on the road, the humidity level, and the tire's temperature. As a result of those factors, an engineer might be fortunate to know the tire's friction coefficient to one significant digit. In that event, it would be meaningless to report numerical values having many digits. Doing so could even be misleading to the extent that another person might interpret the result as being more accurate than it really is.

2.5 AN ERROR OF UNITS ON THE WAY TO MARS

The importance of keeping track of units in engineering calculations was highlighted by the failure of the Mars Climate Orbiter spacecraft in 1999. This spacecraft, which weighed some 1387 pounds and was part of a $125 million planetary exploration program, was designed to be the first orbiting weather satellite for the planet Mars (Figure 2.3). Remote exploration of Mars is an important scientific endeavor. Besides Earth, Mars is the planet in our solar system that has the most hospitable climate, including a landscape that has been shaped by landslides, wind, volcanism, and water. In addition to gathering information about Martian weather, data from the spacecraft would have been useful to help scientists answer such profound questions as: what caused drastic changes in the climate of Mars, and were the conditions necessary for primitive life ever present?

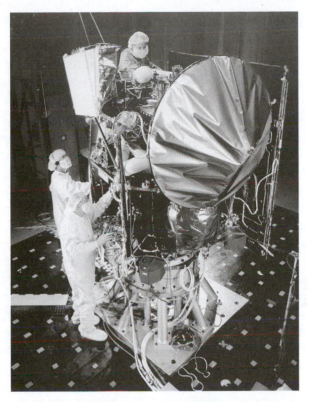

FIGURE 2.3   Technicians and engineers prepare the Mars Climate Orbiter spacecraft prior to its launch.
Source: Reprinted with permission of NASA.

The Mars Climate Orbiter (MCO) was launched aboard a Delta II rocket on December 11, 1998 from Cape Canaveral, Florida. The spacecraft was to arrive at Mars on September 23, 1999, and it was scheduled to complete its primary science mission on December 31, 2004. As the MCO approached the northern hemisphere of Mars, the spacecraft was to fire its main engine for 16 minutes and 23 seconds at a thrust level of 640 N. The engine burn would slow down the spacecraft and place it into an elliptical capture orbit around Mars. During subsequent passes around the planet, thrusters on the MCO would further lower the spacecraft into the more circular orbit needed for its science mission.

However, following the first main engine burn, the National Aeronautics and Space Administration suddenly issued the following statement:

> Mars Climate Orbiter is believed to be lost due to a suspected navigation error. Early this morning at about 2 A.M. Pacific Daylight Time the orbiter fired its main engine to go into orbit around the planet. All the information coming from the spacecraft leading up to that point looked normal. The engine burn began as planned five minutes before the spacecraft passed behind the planet as seen from Earth. Flight controllers did not detect a signal when the spacecraft was expected to come out from behind the planet.

The following day, it was announced that:

> Flight controllers for NASA's Mars Climate Orbiter are planning to abandon the search for the spacecraft at 3 P.M. Pacific Daylight Time today. The team has been using the 70-meter-diameter (230-foot) antennas of the Deep Space Network in an attempt to regain contact with the spacecraft.

What went wrong? Subsequent investigation of the incident and the spacecraft's flight trajectory revealed that during its approach to the planet, the MCO apparently passed only 60 km above the Martian surface, rather than the planned closest approach of between 140 km and 150 km. Flight engineers had considered 85 km as the minimum safe altitude for the spacecraft's flyby. The implication of the unexpectedly low altitude during Mars approach is that either the spacecraft burned up and crashed or it skipped off the atmosphere like a stone on the surface of a lake, became overheated and nonoperational, and began to orbit the sun. Either way, the spacecraft was lost.

NASA conducted a thorough investigation, and the Mars Climate Orbiter Mishap Investigation Board identified eight factors that contributed to the spacecraft's loss. A primary cause was found to be failure to correct an error that occurred when information was transferred between two teams that were collaborating on the spacecraft's operation and navigation. As it turns out, in a segment of the ground-based computer software for navigating the spacecraft, the teams failed to convert a quantity properly between the USCS and SI.

The engineering quantity in question, the engine's impulse, represents the net effect of the rocket engine's thrust during the time of an engine burn. The project's scientists and engineers needed to know the impulse accurately in order to calculate changes in the spacecraft's velocity. Impulse has the units of (force) × (time), and mission specifications called for it to have been given in the units of newton-seconds. Instead, the data were interpreted by one team as being given in the units of pound-seconds. One group of scientists and engineers working on the MCO's navigation thought the impulse was specified in the SI, and the other group understood the numerical values to be given in the USCS. The error resulted in the effect of the main engine burn on the spacecraft's trajectory being off by a factor of roughly 4.45, the conversion factor between newtons and pounds.

This case study emphasizes not only the importance of properly using units in calculations but also the need for good communication channels between technical teams that work together. Proper accounting of units is not necessarily "rocket science," but it is important, and you should begin developing good practices for keeping track of units in your calculations and for reporting them with your numerical answers.

2.6 APPROXIMATION IN ENGINEERING

Engineers nearly always make approximations when they design and solve technical problems. Those approximations are made in order to reduce a real problem into its most basic and essential elements. Approximations are also useful to remove extraneous features that complicate the problem but otherwise have little influence on the final answer. Engineers are comfortable making reasonable approximations so that the systems they analyze are as simple as possible and yet will yield a result that is accurate enough for the task at hand. If the accuracy should need to be increased at some point, then one would need to incorporate more physical phenomena or detailed geometry, and the equations to be solved would likewise become more complicated.

Given the uncertainty present in real systems, it is often necessary for engineers to make "order-of-magnitude" approximations. These are sometimes called "back-of-the-envelope" calculations because they can be performed quickly and informally. For instance, order-of-magnitude approximations can be used to evaluate potential design options for their feasibility or to provide estimates for a structure's weight or the amount of power that is needed. Those quick calculations can be made before significant effort has been put into the details of a design.

One makes order-of-magnitude calculations fully aware of the approximations involved, all the while recognizing that approximations are necessary to reach an answer. The term *order-of-magnitude* implies that the quantities considered in the calculation (and the ultimate answer) are accurate to perhaps a factor of 10 or so. Thus, the calculation might estimate the force carried by a certain bolted connection as being 1000 lb, implying that the force probably isn't as low as 100 lb or as great as 10,000 lb, but it could be 800 lb or 3000 lb. Despite the wide range of possible values, the order-of-magnitude calculation is nevertheless useful because it places a bound on what the answer is expected to be. The estimate also provides a starting point for any subsequent detailed calculations to be performed. Order-of-magnitude estimates are used to double-check and help you decide if an answer is reasonable or if an obvious error has been made. Calculations of this type are educated estimates, admittedly imperfect and imprecise.

Order-of-magnitude estimates are made for the simple and practical reason that in analyzing an engineering problem, one often must start with a blank piece of paper. Engineers need to somehow begin the process of assigning numerical values to dimensions, weights, material properties, and other parameters. You should recognize that the values will be refined in the future as information continues to be gathered, the analysis improved, and the problem better defined. The following examples show some applications of order-of-magnitude calculations and the thought processes behind them.

EXAMPLE 2.5

Commercial jet aircraft have pressurized cabins because they travel at high altitude where the atmosphere is thin. At the cruising altitude of 30,000 feet, the outside atmospheric pressure is only about one-third the sea-level value. The cabin is pressurized to the equivalent of a 10,000 foot altitude, where the air pressure is about 70% that at sea level. Estimate the force that is applied to the door of the aircraft's main cabin caused by this pressure imbalance. Treat the following information as "given" when making your order-of-magnitude estimate: (1) The air pressure at sea level is approximately 15 psi, and (2) the force F on the door is the product of the pressure difference Δp and the door's area A according to the expression $F = A\Delta p$.

SOLUTION

The net pressure on the door is the difference in pressure inside the plane (about 70% sea level) and the outside (about 30% sea level) for a difference of $(0.7 - 0.3)(15 \text{ psi}) = 6$ psi. We estimate the size of the door to be 6 feet by 3 feet, so that its area is about 2600 in^2. In estimating the door's area, we have neglected the facts that the door is not precisely rectangular and that it is curved to blend with the fuselage. The total force acting on the door becomes $(6 \text{ psi})(2600 \text{ in}^2) = 15{,}600 \text{ lb} \approx 15{,}000$ lb. These forces can evidently be quite large, even for seemingly small pressures, when they act over large surfaces. ∎

EXAMPLE 2.6

In an analysis of human-powered machinery, an engineer wants to estimate the amount of power that a person can produce. In particular, can a person who is comfortably riding an exercise bike power a television during the workout? Treat the following information as "given" when making your order-of-magnitude estimate: (1) A typical television consumes 150–200 W of electrical power, and (2) a mathematical expression for power P is

$$P = \frac{Fd}{\Delta t}$$

where F is the magnitude of a force, d is the distance over which it acts, and Δt is the time interval during which the force is applied.

SOLUTION

To estimate power output, we make a comparison with the rate at which a person can comfortably climb a flight of stairs. It seems reasonable to say that a flight of stairs, about a 3 m rise, can be covered in 10 s by a person weighing 700 N. In that case, the power becomes $(700 \text{ N})(3 \text{ m})/(10 \text{ s}) = 210$ W. The useful work that a person can produce is therefore about 200 W. However, the generator that would be used to convert the mechanical work into electricity will not be perfectly efficient, and it might have a conversion efficiency of only 20%. Thus, we would expect that only about $(200 \text{ W})(0.20) = 40$ W or so of electricity could be produced, an amount insufficient to power a full-sized television. ∎

2.7 PROBLEM-SOLVING METHODOLOGY

Engineers are very organized people. Over the years and across many industries and technical fields, such as those described in Chapter 1, engineers have developed a well-deserved reputation for being clear thinkers who pay attention to detail and who get answers right. That respect has been earned, and based on it, the public gives its trust and confidence to the products that engineers design and build.

Engineers expect one another to present calculations, test results, and technical work in a clear, convincing, and well-documented manner. In beginning your study of mechanical engineering, you should start to develop organizational and problem-solving skills that meet the standards of what other engineers will expect from you. There are two reasons behind this advice. First, by following a systematic approach to solving problems, you will reduce the chance of having common but preventable mistakes creep into your solutions. Errors involving algebra, dimensions, units, conversion factors, and misinterpretation of a problem statement are preventable. By paying attention to detail, you can keep those types of errors to a minimum. Second, your good procedure and format for solving engineering problems will help others interpret and understand your work. Engineers do not work in isolation; their work is read, critiqued, interpreted, modified, verified, and built upon by their co-workers, clients, customers, and supervisors. If others can't understand your work, or they don't have confidence that it was double-checked and developed carefully, it will not be well regarded, to say the least. Proper problem-solving methodology is important for you to get the correct answer and to convincingly communicate the result to others.

Now is the right time for you to start developing these skills. For the purpose of solving the problems in this textbook, the following guidelines will be helpful:

1. **Make a clean start.** Start each problem at the top of a new page, even if the solution will be less than one page in length. Separating problems in this manner organizes the work and makes it easier for others to locate and read each of your solutions. Only write on one side of the paper. This habit also makes it easier for others to quickly find, review, and read your work.

2. **Draw.** Include a drawing of the system that is being analyzed, labels for the major components, and relevant dimensions. Engineers tend to think and learn visually, and your drawings and graphs are the best way to convey complex information quickly. An important first step in solving nearly every engineering problem is to represent the problem graphically.

3. **Givens and unknowns.** Write a short summary of the problem's givens and unknowns, perhaps one or two sentences in length. By being clear as to the objective of the problem, you will disregard extraneous information and focus on solving the problem in a more direct manner.

4. **Think first, then write.** Think about the problem before you start crunching numbers and putting pencil to paper. Make a short summary in words to explain the

general approach that you will take, and list the major concepts, equations, and assumptions that you expect to use.

5. **Be coordinated.** Draw the coordinate system being used in order to show the directions and positive sign conventions that you are adopting. If you write "$F_x = 25$ lb" in a solution, for instance, but you do not indicate in a drawing what you consider to be the x direction, the solution will mean little to someone reviewing the work. By the same token, if you should analyze a geartrain or transmission and report the speed of a shaft to be 750 rpm (revolutions per minute), you will also need to indicate the direction of the shaft's rotation.

6. **Neatness counts.** It is not sufficient that your calculation makes sense to you. It must also make sense to others who want to read, but not decipher, your work. If another engineer can't review and quickly understand your work, it will be ignored or viewed as being confusing, incomplete, or inaccurate. Take the time to write clearly and document your steps, for that is a sign of a conscientious and professional effort.

7. **Units.** At each step in your calculation, and in the final answer, be sure to include the units with each numerical value. A number without a unit is meaningless, just as a unit is meaningless without a numerical value assigned to it.

8. **Significant figures.** Check the number of significant digits in your answer and be sure that the solution is not presented as being more accurate than it really is.

9. **Box it.** Underline, circle, or box your final result so that there is no ambiguity about the answer that you are reporting.

10. **Interpret your answer.** Explain what your answer means from a physical standpoint. How reasonable does the result seem to you? Does the order-of-magnitude of the result make sense based on your experience, judgment, and intuition? Use this test as a "commonsense" double-check.

SUMMARY

Engineers are often described as being "can-do" people with excellent problem-solving skills. In this chapter we have discussed some of the fundamental tools and skills that mechanical engineers use when they answer technical questions. Numerical values, the USCS and SI systems, unit conversions, dimensional consistency, significant digits, and order-of-magnitude approximations are everyday issues for engineers. Each quantity in mechanical engineering has two components—a numerical value and a unit—and it is meaningless to report one without the other. Mechanical engineers need to be clear about those numerical values and units when they perform calculations, document results, and communicate their findings to others through written reports and oral presentations. By following the consistent problem-solving guidelines developed in this chapter, you will be able to approach engineering problems in a systematic manner and be confident of the accuracy of your solutions.

SELF-STUDY AND REVIEW

1. What are the base units in the USCS and SI?
2. What are examples of derived units in the USCS and SI?
3. How are mass and force treated in the USCS and SI?
4. One pound is equivalent to approximately how many newtons?
5. One meter is equivalent to approximately how many feet?
6. One inch is equivalent to approximately how many millimeters?
7. One gallon is equivalent to approximately how many liters?
8. How should you decide the number of significant digits to retain in a calculation and report in your final answer?
9. Give an example of the process of making order-of-magnitude approximations.
10. Summarize the major steps that should be followed when solving problems to clearly document your work and to catch otherwise avoidable mistakes.

PROBLEMS

1. Express your weight in the units lb and N and your mass in the units slug and kg.
2. Express your height in the units in., ft, and m.
3. One U.S. gallon is equivalent to 0.1337 ft^3, 1 foot is equivalent to 0.3048 m, and 1000 liters is equivalent to 1 m^3. By using those definitions, determine the conversion factor between gallons and liters.
4. A passenger automobile is advertised as having a fuel economy rating of 29 mi/gal for highway driving. Express the rating in the units of km/L.
5. (a) How many horsepower does a 100 W household lightbulb consume? (b) How many kW does a 5 hp lawn mower engine produce?
6. Calculate various fuel quantities for Flight 143. The plane already had 7682 L of fuel on board prior to the flight, and the tanks were to be filled so that a total of 22,300 kg were present at takeoff. (a) Using the incorrect conversion factor of 1.77 kg/L, calculate in units of kg the amount of fuel that was added to the plane. (b) Using the correct factor of 1.77 lb/L, calculate in units of kg the amount of fuel that should have been added. (c)

By what percent would the plane have been underfueled for its journey? Be sure to distinguish between weight and mass quantities in your calculations.

7. Printed on the side of a tire on an all-wheel-drive sport utility wagon is the warning "Do not inflate above 44 psi," where psi is the abbreviation for the pressure unit lb/in^2. Express the tire's maximum pressure rating in (a) the USCS unit of lb/ft^2 (psf) and (b) the SI unit of kPa.
8. The amount of power transmitted by sunlight depends on latitude and the surface area of the solar collector. On a clear day at a certain northern latitude, 0.6 kW/m^2 of solar power strikes the ground. Express that value in the alternative USCS units of (a) (ft·lb/s)/ft^2 and (b) hp/ft^2.
9. The property of a fluid called *viscosity* is related to its internal friction and resistance to being deformed. The viscosity of water, for instance, is less than that of molasses and honey, just as the viscosity of light motor oil is less than that of grease. A unit used in mechanical engineering to describe

viscosity is called the *poise*, named after the physiologist Jean Louis Poiseuille who performed early experiments in fluid mechanics. The unit is defined by 1 poise = 0.1 N·s/m^2. Show that 1 poise is also equivalent to 1 g/(cm·s).

10. Referring to the description in Problem 9, and given that the viscosity of a certain engine oil is 0.25 kg/(m·s), determine the value in the units (a) poise and (b) slug/(ft·s).

11. Referring to the description in Problem 9, if the viscosity of water is 0.01 poise, determine the value in terms of the units (a) slug/(ft·s) and (b) kg/(m·s).

12. The fuel efficiency of an aircraft's jet engines is described by the thrust-specific fuel consumption, or TSFC. The TSFC measures the rate of fuel consumption (mass of fuel burned per unit time) relative to the thrust (force) that the engine produces. In that manner, just because an engine might consume more fuel per unit time than a second engine, it would not necessarily be more inefficient if it also produced more thrust to power the plane. The TSFC for an early hydrogen-fueled jet engine was 0.082 (kg/h)/N. Express that value in the USCS units of (slug/s)/lb.

13. An automobile engine is advertised as producing a peak power of 118 hp (at an engine speed of 4000 rpm) and a peak torque of 186 ft·lb (at 2500 rpm). Express those performance ratings in the SI units of kW and N·m.

14. For the exercise of Example 2.3, express the sideways deflection of the tip in the units of mils (defined in Table 2.2) when the various quantities are instead known in the USCS. Use the values $F = 75$ lb, $L = 3$ in., $d = \frac{3}{16}$ in., and $E = 30 \times 10^6$ psi.

15. Heat, ΔQ, with the SI unit of joule (J), is the quantity in mechanical engineering that describes the transit of energy from one location to another. The equation for the flow of heat during the time interval Δt through an insulated wall is

$$\Delta Q = \frac{\kappa A \Delta t}{L}(T_h - T_l)$$

where κ is the thermal conductivity of the material from which the wall is made, A and L are the wall's area and thickness, and $T_h - T_l$ is the difference (in degrees Celsius) between the high- and low-temperature sides of the wall. By using the principle of dimensional consistency, what is the correct unit for thermal conductivity in the SI? The Greek letter kappa (κ) is a conventional mathematical symbol used for thermal conductivity. Appendix A summarizes the names and symbols of Greek letters.

16. Convection is the process by which warm air rises and cooler air falls. The Prandtl number, Pr, is used when mechanical engineers analyze certain heat transfer and convection processes. It is defined by the equation

$$Pr = \frac{\mu c_p}{\kappa}$$

where c_p is a property of the fluid called the *specific heat* having the SI units kJ/(kg·°C); μ is the viscosity as discussed in Problem 9; and κ is the thermal conductivity as discussed in Problem 15. Show that Pr is a dimensionless number. The Greek letters mu (μ) and kappa (κ) are conventional mathematical symbols used for viscosity and thermal conductivity. Appendix A summarizes the names and symbols of Greek letters.

17. Referring to Problem 16 and Table 2.2, if the units for c_p and μ are Btu/(slug·°F) and slug/(ft·h), respectively, what must be the USCS units of thermal conductivity in the definition of Pr?

18. Some scientists believe that the collision of one or more large asteroids with the Earth was responsible for the extinction of the dinosaurs. The unit of kiloton is used to

describe the energy released during large explosions. It was originally defined as the explosive capability of 1000 tons of trinitrotoluene (TNT) high explosive. Because that expression can be imprecise depending on the explosive's exact chemical composition, the kiloton has been subsequently redefined as the equivalent of 4.186×10^{12} J. In the units of kiloton, calculate the kinetic energy of an asteroid that has the size (box-shaped, $13 \times 13 \times 33$ km) and composition (density, 2.4 g/cm^3) of our solar system's asteroid Eros. Kinetic energy is defined by

$$\Delta U = \frac{1}{2}mv^2$$

where m is the object's mass and v is its speed. Objects passing through the inner solar system generally have speeds in the range of 20 km/s.

19. A structure known as a cantilever beam is clamped at one end but free at the other, analogous to a diving board that supports a swimmer standing on it. Using the following procedure, conduct an experiment to measure how the cantilever beam bends. In your answer, report only those significant digits that you know reliably.

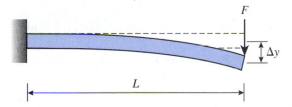

- Make a small tabletop test stand to measure the deflection of a plastic drinking straw (your cantilever beam) that bends as a force F is applied to the free end. Push one end of the straw over the end of a pencil, and then clamp the pencil to a desk or table. You can also use a ruler, chopstick, or a similar component as the cantilever beam itself. Sketch and describe your apparatus, and measure the length L.
- Apply weights to the end of the cantilever beam, and measure the tip's deflection Δy using a ruler. Repeat the measurement with at least a half-dozen different weights to fully describe the beam's force–deflection relationship. Penny coins can be used as weights; one penny weighs approximately 30 mN. Make a table to present your data.
- Next draw a graph of the data. Show tip deflection on the abscissa and weight on the ordinate, and be sure to label the axes with the units for those variables.
- Draw a best-fit line through the data points on your graph. In principle, the deflection of the tip should be proportional to the applied force. Do you find this to be the case? The slope of the line is called the *stiffness*. Express the stiffness of the cantilever beam either in the units lb/in. or N/m.

20. Perform measurements as described in Problem 19 for cantilever beams of several different lengths. Can you show experimentally that for a given force F, the deflection of the cantilever's tip is proportional to the cube of its length? As in Problem 19, present your results in a table and a graph, and report only those significant digits that you know reliably.

21. Estimate the force acting on a passenger window in a commercial jet aircraft due to air pressure differential.

22. Give numerical values for order-of-magnitude estimates for the following quantities. Explain and justify the reasonableness of the assumptions and approximations that you need to make.

- The number of cars that pass through an intersection of two busy streets during the evening commute on a typical workday.
- The number of bricks that form the exterior of a large building on a university campus.

- The volume of concrete in the sidewalks on a university campus.

23. Repeat the exercise of Problem 22 for the following systems:

    - The number of leaves on a mature maple or oak tree.
    - The number of gallons of water in an Olympic-sized swimming pool.
    - The number of blades of grass on a natural-turf football field.

24. The Space Shuttle orbiter takes about 90 minutes to complete one trip around the Earth. Estimate the spacecraft's orbital velocity, in the units of mph. Make the approximation that the altitude of the spacecraft (approximately 125 miles) is small when compared to the radius of the Earth (approximately 3950 miles).

25. Estimate the size of a square parcel of land that is needed for an airport's 5000-car parking lot. Include space for the access roadways.

26. Make an order-of-magnitude estimate of the number of gallons of gasoline that are consumed by automobiles in the United States each day. Explain and justify the reasonableness of the assumptions and approximations that you make.

27. An automobile assembly plant produces 400 vehicles per day. Make an order-of-magnitude estimate for the weight of the steel needed to make those vehicles. Explain and justify the reasonableness of the assumptions and approximations that you make.

28. Think of some quantity that you encounter in your day-to-day life for which it would be difficult (or impossible) to obtain a highly accurate numerical value but for which an order-of-magnitude approximation can be made. Describe the quantity, make the approximation, and explain and justify the reasonableness of the assumptions and approximations that you need to make.

REFERENCES

Banks, P., "The Crash of Flight 143," *ChemMatters*, American Chemical Society, October 1996, p. 12.

Hoffer, W., and Hoffer, M. M., *Freefall: A True Story*, St. Martin's Press, New York, 1989.

Mars Climate Orbiter Mishap Investigation Board Phase I Report, NASA, November 10, 1999.

Press Release, "NASA's Mars Climate Orbiter Believed To Be Lost," Media Relations Office, Jet Propulsion Laboratory, September 23, 1999.

Press Release, "Mars Climate Orbiter Mission Status," Media Relations Office, Jet Propulsion Laboratory, September 24, 1999.

Press Release, "Mars Climate Orbiter Failure Board Releases Report, Numerous NASA Actions Underway In Response," NASA, November 10, 1999.

3 Machine Components and Tools

3.1 OVERVIEW

As introduced in Chapter 1, the primary objective of mechanical engineers is to make machines that work, that are useful, and that improve society in some way. Machines in general, and power transmission equipment in particular, are often constructed of standardized components or "building blocks." Just as an electrical engineer might select off-the-shelf resistors, capacitors, and transistors as the elements of a circuit, mechanical engineers have good intuition for specifying the different available types of bearings, shafts, gears, belts, and other components in their designs. Once the details of a machine have been worked out and the components for it have been selected, the product then needs to be built. Mechanical engineers also devise manufacturing and assembly processes, and at the conclusion of this chapter we will introduce several common machine tools.

A solid appreciation of machine components and tools is valuable at three levels. First, and from our primary standpoint in this textbook, a working knowledge of this hardware is important for you to develop a technical vocabulary. Mechanical engineering has its own precise language, and specialized terms are used to describe the construction and operation of machines. To communicate effectively with other engineers, you will need to learn, adopt, and share that language. Second, having such a background is necessary to select the proper component for an application and to specify the appropriate machine tool for a production process. Of course, it is not possible to list and describe every machine and component that embodies mechanical engineering principles, and that is not our intention in this chapter. However, by examining just a few machine components and tools, you will develop a growing appreciation for practical machinery issues. Finally, this chapter is a good place for you to begin the journey of understanding the inner workings of mechanical hardware. It's intellectually healthy for you to be curious about machines, wonder how they were made, dissect them, and think about how they could have been made differently or better (Figure 3.1).

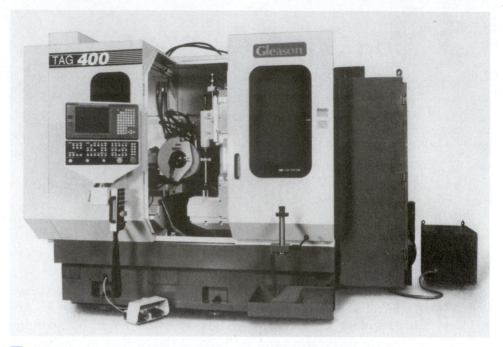

After completing this chapter you should be able to:

- Describe the structure and characteristics of ball, straight roller, tapered roller, and thrust bearings.
- Understand the purpose of flexible shaft couplings.
- Describe the structure and characteristics of spur, rack, bevel, helical, and worm gears.
- Describe the differences between v-belts and timing belts.
- Explain the circumstances in which one type of bearing, gear, or drive belt would be selected for use instead of another.
- Describe how drill presses, band saws, lathes, and mills are used in machine shops for manufacturing.

3.2 ROLLING ELEMENT BEARINGS

Bearings are used to hold shafts that must rotate relative to supports that are fixed—for instance, the housing of a motor, gearbox, or transmission. A wide variety of bearings exist, and each particular type is appropriate for a different set of installation and operating conditions. Although a ball bearing might be the best choice for one application, it might not have enough load-carrying capability for another. The common feature of all bearings is that they enable shafts to rotate with little resistance while being well supported against forces acting in other directions.

Bearings are classified into two broad groups, called *rolling contact* and *journal*. In this chapter, we discuss only rolling contact bearings; these typically comprise an inner race; an outer race; rolling elements in the form of balls, cylinders, or cones; and a separator that prevents the rolling elements from rubbing against one another. Rolling contact bearings are ubiquitous in machine design, and they are found in applications as diverse as bicycle wheels, robotic joints, and automobile transmissions. Journal bearings, on the other hand, have no rolling elements. Instead, the shaft simply rotates within a polished sleeve that is lubricated by oil or another fluid. Just as the puck on an air hockey table slides smoothly over a thin film of air, the shaft in a journal bearing is supported by a thin film of oil. Also ubiquitous, journal bearings support crankshafts in internal combustion engines and shafts inside pumps and compressors.

Journal bearing
- - - - - - - - - - - - - - →

A sample installation of a rolling contact bearing is illustrated in Figure 3.2. The shaft and the bearing's inner race rotate together, while the outer race and the transmission's housing are stationary. When a shaft is supported in this way, the bearing's outer race will fit tightly into a matching circular recess that is formed in the housing. As the shaft in Figure 3.2 turns and transmits power, the bearing could be subjected to forces directed either along the shaft (called a *thrust force*) or perpendicular to it (a *radial force*). An engineer will make a decision on the type of bearing to be used in a machine depending on whether thrust forces, radial forces, or

Thrust force, radial force
- - - - - - - - - - - - - - →

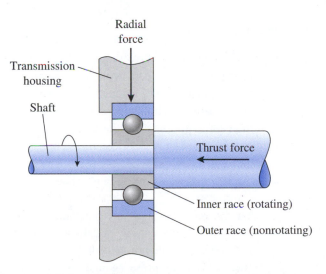

■ **FIGURE 3.2** A sample installation of a bearing in a transmission's housing.

■ **FIGURE 3.3** Pillow block bearings can be used to support a power transmission shaft.

some combination of the two act on the bearing. Thus, engineers apply the properties of force systems (as described in Chapter 4) when they analyze and select bearings and other machine components.

Rolling contact bearings can also support shafts that are mounted on the outside of a frame or structure. In those cases, it may not be possible to press the bearing into a recess such as that shown in Figure 3.2. One solution for such situations is to use a *Pillow block* pillow block mount, as depicted in Figure 3.3. The bearing itself is contained within the block, which in turn can simply be bolted to a suitable surface. Because of the clearances that are present in the pillow block's bolt holes, this type of mounting arrangement generally does not provide close tolerance for positioning and aligning the shaft.

Ball Bearings

The most common type of rolling element bearing is the ball bearing, which *Inner and outer* incorporates hardened, precision-ground steel spheres. Figure 3.4 depicts the *races* major elements of a standard ball bearing: the inner race, outer race, balls, and separator. The inner and outer races are the bearing's connections to the (rotating) shaft and the (stationary) housing. The separator (which can also be called the *cage* or *retainer*) keeps the balls evenly spaced around the bearing's perimeter and prevents them from contacting one another. Otherwise, if *Separator* the bearing was used at high speed or subjected to large forces, friction could cause it to overheat and become damaged. In some cases, the gap between the inner and outer races will be sealed by a rubber or plastic *Seals* ring to keep grease in the bearing and dirt and debris out of it. Thus, bearings can either be sealed or unsealed.

In principle, the balls press against the inner and outer races of the bearing at single points. The force that each ball transfers between the inner and outer races is

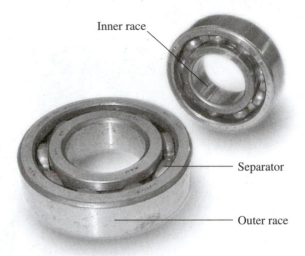

Inner race

Separator

Outer race

■ **FIGURE 3.4** Elements of a standard ball bearing.

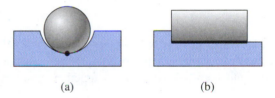

(a)　　　　　　(b)

■ **FIGURE 3.5** Detail of (a) point contact between a spherical ball and the raceways of a bearing, and (b) line contact in a straight roller bearing.

concentrated on those surfaces in an intense and relatively sharp manner, as indicated in Figure 3.5(a). If the forces between the rolling elements and races were instead spread over a larger area, we might expect the bearing to last longer. Rolling contact bearings that incorporate cylindrical rollers or tapered cones in place of spherical balls are one solution for more evenly distributing forces, and we discuss those types of bearings next.

Straight Roller Bearings

While ball bearings may be inexpensive, they have a relatively modest capacity for carrying forces because of the point contacts present between the balls and races. As shown in Figure 3.5(b), cylindrical rollers can also be used in a rolling element bearing

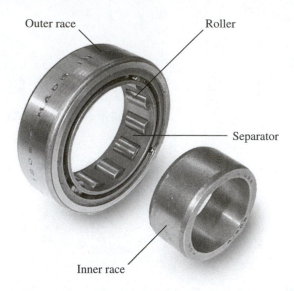

■ **FIGURE 3.6** Structure of a straight roller bearing. The inner race has been removed in order to show the rollers and separator.

in order to better distribute forces over the bearing's races. Figure 3.6 illustrates the structure of a straight roller bearing. If you place a few pens or pencils between your hands, and then rub your hands together, you have the essence of a straight roller bearing.

In Figure 3.6, the centerlines of the rollers and the races are parallel to one another, and a separator is again used to space rollers around the bearing's circumference. Straight roller bearings are used in machinery when large radial forces must be carried, but they are not well suited for supporting thrust loads.

Tapered Roller Bearings

Instead of having spheres or cylinders roll on the races, a tapered roller bearing uses rollers that are shaped like truncated cones, as seen in Figure 3.7. The conical rollers all have the same taper angle and their centerlines intersect at a single point on the shaft, located some distance away from the bearing itself. Separators in the bearings of Figures 3.7 and 3.8 prevent friction from roller-to-roller contact and better distribute forces around the bearing. As was the case for straight roller bearings, the rollers of a tapered roller bearing contact the outer race (which is often called the *cup*) and the inner race (the *cone*) along lines, rather than points, in order to lower wear. The lifetime of straight and tapered roller bearings is generally further improved by giving the rollers

Inner race

Separator

Roller cones

Outer race

■ **FIGURE 3.7** Structure of a tapered roller bearing.

■ **FIGURE 3.8** A type of tapered roller bearing that is widely used in automotive front wheels, differentials, and machine tool spindles.
Source: Reprinted with permission of The Timken Company.

a slight crown or barrel shape, a feature again intended to smooth out the transfer of forces between the races.

Although they may not be as familiar to you as ball bearings, tapered roller bearings are quite common in mechanical engineering. They are intended for applications where both thrust and radial forces are present—the wheel bearings in an automobile

FIGURE 3.9 A four-row tapered roller bearing used in sheet metal rolling mills and other heavy-duty applications.
Source: Reprinted with permission of The Timken Company.

are a prime example. Wheel bearings must support both the weight of the vehicle (which is a radial force between the bearing and the wheel's shaft) and the cornering force generated during turns (which is a thrust force).

Tapered roller bearings are regarded as having an intermediate cost, a high capacity for radial forces, and a moderate capacity for thrust forces. In other high-load and heavy-duty applications, matched tapered roller bearings are combined into two-row or four-row sets, such as those shown in Figure 3.9.

Thrust Roller Bearings

While straight roller bearings are well suited for supporting loads that are directed mostly radially, and tapered roller bearings can support a combination of radial and thrust forces, thrust roller bearings carry loads that are directed primarily along a shaft's centerline. One type of thrust bearing is shown in Figure 3.10. The rolling elements in this type of bearing are cylinders having a slight barrel shape. In contrast to the straight roller bearing of Figure 3.6, these rollers are placed radially relative to the shaft. Thrust bearings are appropriate for such applications as a turntable that must support the dead weight of cargo but which also needs to turn freely. The thrust bearing shown in Figure 3.11 uses tapered rollers.

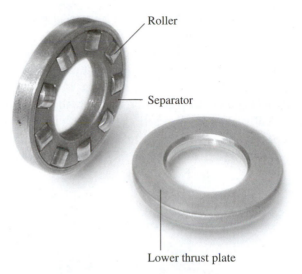

Roller

Separator

Lower thrust plate

■ **FIGURE 3.10**   Structure of a thrust roller bearing.

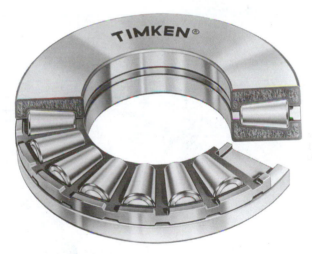

■ **FIGURE 3.11**   A thrust bearing comprised of tapered rollers, two thrust races, and a separator.
Source: Reprinted with permission of The Timken Company.

3.3 FLEXIBLE SHAFT COUPLINGS

When the output shaft from one machine is connected to the input shaft of another, you might initially expect that the shafts would always be rigidly connected. From a practical standpoint, however, it is difficult to align two shafts precisely, and any (even slight) angular or offset misalignment can cause the bearings to become over loaded and damaged.

Consider the arrangement in which the shaft of an electric motor will be connected to another shaft that serves as the input to a gearbox. The electric motor in this case has its own shaft that is mounted on bearings internal to the motor. Our conceptual configuration is shown in Figure 3.12(a), where the shafts of the motor and the transmission are ideally aligned. In practice, however, small but significant imperfections due to manufacturing and assembly tolerances will be unavoidable. In Figure 3.12(b), if the two shafts were rigidly connected, the misalignment would cause the shafts to bend. In

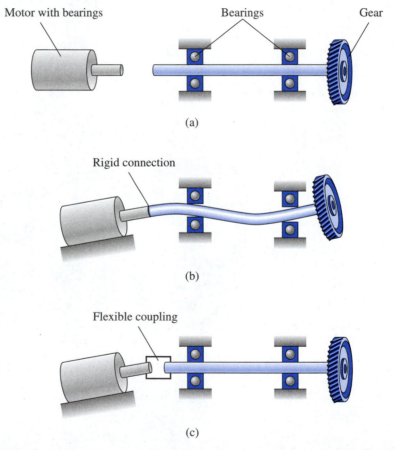

■ **FIGURE 3.12**   The shaft of an electric motor is connected to a geartrain. (a) The two shafts are perfectly (but unrealistically) aligned. (b) When the shafts are slightly misaligned and rigidly connected, the shaft bends and the bearings are subjected to large forces. (c) A flexible coupling placed between the two shafts accommodates misalignment while still transmitting rotation. The amount of misalignment and bending are exaggerated for clarity.

■ **FIGURE 3.13**   Samples of flexible shaft couplings.
Source: Reprinted with permission of W. M. Berg, Incorporated.

turn, the bearings would be subjected to relatively large forces, and they would not be expected to have a long service life.

   To overcome this problem, a flexible coupling can be used to connect the two shafts. Several types are shown in Figure 3.13. These couplings effectively transmit rotation, and at the same time they are relatively compliant with respect to sideways bending. In short, flexible couplings accommodate shafts that need to be connected but that might be slightly misaligned, and in so doing they reduce forces acting on the drivetrain's bearings.

3.4   GEAR TYPES AND TERMINOLOGY

Gears, and the belt and chain drives that are described in Section 3.5, transmit rotation, torque, and power between shafts. Geartrains can be used to increase rotation speed but decrease torque, to keep speed and torque constant, or to reduce rotation speed but increase torque. Mechanisms incorporating gears are common in the design of machinery, and they have applications as diverse as electric can openers and helicopter transmissions. The mathematical analysis of geartrains and other machinery is postponed until Chapter 7, at which point we will have set a foundation in place for analyzing force, torque, and power. In this section, our objective is simply to explore various types of gears with an emphasis on their characteristics and the terminology used to describe them.

The shape of a gear's tooth is mathematically defined and precisely machined according to codes and standards that have been established by industry. The American Gear Manufacturers Association, for instance, has developed guidelines for standardizing the design and production of gears. Mechanical engineers are able to purchase pairs of loose gears directly from gear manufacturers and suppliers, or they can obtain prefabricated gearboxes and transmissions suitable for the task at hand. There are exceptions for circumstances where standard gears might not offer sufficient performance (such as low noise and vibration), and in those cases, contract machine shops can custom-produce gears. In the majority of machine design situations, however, gears and gearboxes are selected as "off-the-shelf" components.

No single "best" type of gear exists, and each variant is well suited to a different application. In the following subsections, the gears known as spur, rack, bevel, helical, and worm are described. The type that an engineer ultimately chooses for a product will reflect a balance between expense and the task that the gear is expected to perform.

Spur Gears

External gear

Internal gear, ring gear

Gearset

Pinion, mesh point

Spur gears are the simplest type of practical engineering-grade gear. As shown in Figure 3.14, spur gears are cut from cylindrical blanks and their teeth have faces that are oriented parallel to the shaft on which the gear is mounted. For the external gears of Figure 3.15(a), the teeth are formed on the outside of the cylinder; conversely, for an internal or ring gear, the teeth are located on the inside (Figure 3.15(b)). When two gears having complementary teeth engage and motion is transmitted from one shaft to another, the two gears are said to form a *gearset*. Figure 3.16 depicts a spur gearset and some of the terminology used to describe the geometry of the teeth. By convention, the smaller (driving) gear is called the *pinion*, and the other (driven) one is simply called the *gear*. The pinion and gear are said to *mesh* at the point where the teeth approach, contact one another, and then separate.

Conceptually, the pinion and gear are regarded as two cylinders that are pressed against one another and that smoothly roll together, in contrast to a collection of many discrete teeth that are continuously contacting, engaging, and disengaging. As illustrated in Figure 3.17, the cylinders roll on the outside of one another for two external gears, or one can roll within the other (one external and one internal

Pitch radius

Pitch circle

gear). Referring to Figure 3.16, the effective radius of a spur gear, which is also the radius of its conceptual rolling cylinder, is called the *pitch radius*, r. Continuous contact between the pinion and gear is imagined to take place on the two pitch circles. The pitch radius is not the distance from the gear's center to either the top or bottom lands of a tooth. Instead, r is simply the radius that an equivalent cylinder would have if it rotated at the same speed as the pinion or gear.

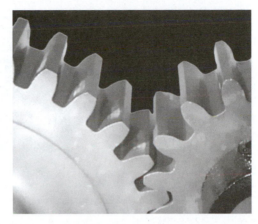

■ **FIGURE 3.14**   Close-up view of two spur gears in mesh.

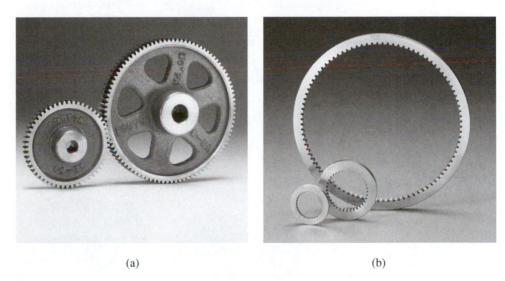

(a)   (b)

■ **FIGURE 3.15**   (a) Two external spur gears in mesh and (b) internal or ring gears of several sizes.

Source: Reprinted with permission of Boston Gear Company.

The thickness of a tooth, and the spacing between adjacent teeth, are each measured along the gear's pitch circle. The tooth-to-tooth spacing will be slightly larger than the tooth's thickness itself in order to prevent the teeth from binding against one another as the pinion and gear rotate. On the other hand, if the space between the teeth is too large, the free play could cause undesirable rattle, vibration, and speed fluctuations. In the USCS, the proximity of teeth to one another is measured by the diameter pitch

$$p = \frac{N}{2r} \tag{3.1}$$

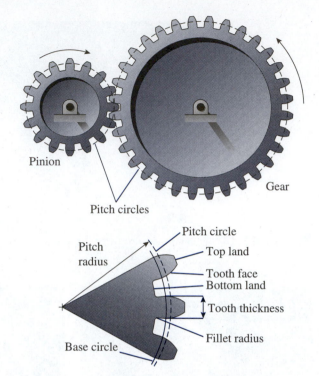

FIGURE 3.16   Terminology for a spur gearset and geometry of its teeth.

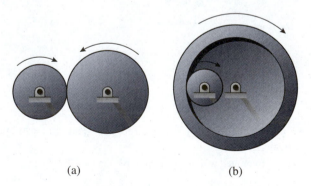

(a)   (b)

FIGURE 3.17   Configurations of gearsets having (a) two external gears and (b) one external and one internal gear. In each case, the rotations are analogous to two cylinders that roll on one another.

Diametral pitch
- - - - - - - - - - - - - - - - →

where N is the number of teeth on the gear. The diametral pitch therefore has the units of teeth per inch of diameter. In order for a pinion and gear to be compatible, they must have the same value of p or the gearset will bind as the shafts begin to rotate. The product catalogs supplied by gear manufacturers generally group compatible gears according to their diametral pitch. By convention, values of $p < 20$ teeth/in. are regarded as being "coarse" gears, and gears with $p \geq 20$ teeth/in. are said to be "fine" pitched. Because r and N are proportional in Equation (3.1) for gears of the same diametral pitch, if a gear's radius is doubled, the number of teeth will likewise double. In the SI, the spacing between teeth is instead measured by a quantity called the module m, which is the reciprocal of the diametral pitch:

Module
------------------->

$$m = \frac{2r}{N} \tag{3.2}$$

and it has the units of millimeters.

In principle, a pinion and gear having teeth with any complementary shape are capable of transmitting rotation between their shafts. In the past, water- and wind-powered mills, for instance, often used "gears" with teeth that were nothing more than wooden pegs attached to a rotating disk. As you might imagine, however, such cog teeth did not provide smooth and efficient operation. With arbitrarily shaped gear teeth, if the (driving) pinion were to turn at a constant speed, the (driven) output gear would be observed to rotate with slight variations in its speed as the teeth engage and disengage. Although perhaps satisfactory for an eighteenth-century grain mill, such changes in output speed are unacceptable for modern high-speed machinery, and they would cause noise, vibration, wear, and inefficiency.

Figure 3.18 shows a sector of a spur gearset during three stages of its rotation. As the pinion and gear mesh, the two teeth roll on the surface of one another and their point of contact moves from one side of the pitch circles to the other. Because of the special mathematical shape of the teeth on a spur gear, however, motion is smoothly transferred between the pinion and gear. The cross-sectional shape of the tooth is called the *involute profile*, and it compensates for the fact that the tooth-to-tooth contact point moves during meshing. The term *involute* is synonymous with *intricate*, and this shape ensures that the (output) gear will turn at a constant speed if the (input) pinion does.

Involute profile
------------------->

You can sketch the involute gear tooth profile by drawing a circle (called the *base circle*) of any radius that you choose, and then laying a short length of string along the circle. Hold one end of the string fixed, and gradually unwrap the string's other end, tangent to the circle. Mark down the curve that the string's free end follows, and you will have drawn the involute's shape. This geometric construction is illustrated in Figure 3.19, where the base circle is slightly smaller than the pitch circle. The mathematical characteristics of the involute profile establish what has become known as the fundamental property of gearsets, and this principle enables engineers to view them as cylinders rolling on one another: For spur gears with involute-shaped teeth, if one gear turns at a constant speed, so will the other.

Base circle
------------------->

Fundamental property of gearsets
------------------->

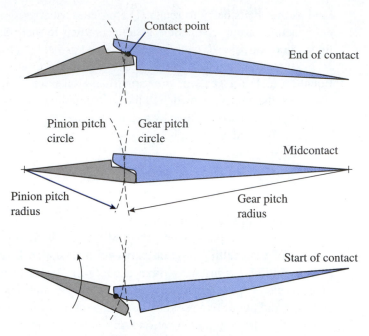

FIGURE 3.18   A pair of engaging and disengaging spur gear teeth.

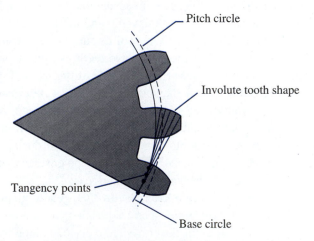

FIGURE 3.19   Drawing the involute tooth profile of a spur gear by unwrapping a string segment from the base circle.

Rack and Pinion

Gears are sometimes used to convert the rotational motion of a shaft into the straight-line motion of a slider and vice versa. The rack and pinion form the limiting case of a gearset in which the gear has an infinite radius. This configuration is shown in Figure 3.20, where the pinion meshes with a straight toothed rack. With the center point of the pinion being fixed, the rack will move horizontally as the pinion rotates, leftward in the figure as the pinion turns clockwise and vice versa. The rack itself (Figure 3.21) can be supported by rollers, or it might slide on a hard surface lubricated with oil. Racks and pinions are often used in the mechanism for steering an automobile's front wheels, and they also

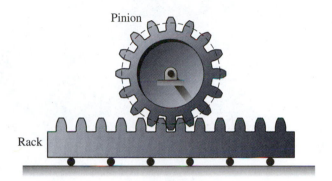

Pinion

Rack

■ **FIGURE 3.20** A rack and pinion mechanism.

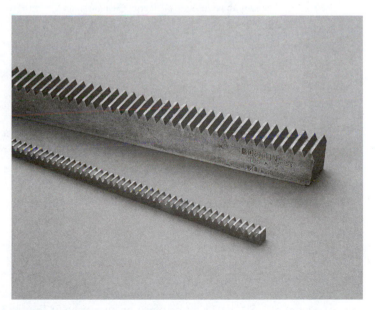

■ **FIGURE 3.21** Samples of gear racks.
Source: Reprinted with permission of Boston Gear Company.

■ **FIGURE 3.22** A collection of different spur gears, including external gears, internal or ring gears, and racks.
Source: Reprinted with permission of Boston Gear Company.

positioned read/write heads above magnetic disks in vintage computer storage drives. A collection of spur gear and rack components is shown in Figure 3.22.

Bevel Gears

Whereas the teeth of spur gears are arranged on a cylinder, a bevel gear is produced by instead forming teeth on a truncated cone. Figure 3.23 is a photograph and cross-sectional drawing of a bevel gearset; you can see how it is able to redirect the shafts by 90°. Bevel gears (Figure 3.24) are appropriate for applications in which two shafts must be connected at a right angle and where extensions to the shaft centerlines would intersect.

Helical Gears

Because the teeth on the spur gears of Figure 3.14 are straight with faces parallel to their shafts, as two teeth approach one another, they make contact along the full width of each tooth. Similarly, in the meshing sequence of Figure 3.18, the teeth separate and lose contact along the tooth's entire width at once. Those relatively sudden engagements

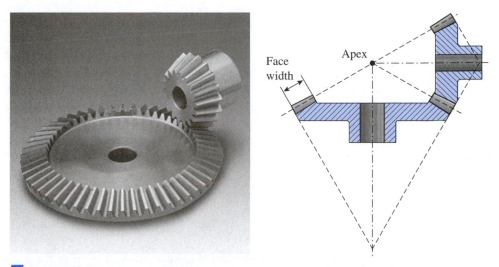

■ **FIGURE 3.23**   Two bevel gears in mesh.
Source: Photograph reprinted with permission of Boston Gear Company.

■ **FIGURE 3.24**   A representative collection of bevel gears, some having straight and others having spiral-shaped teeth.
Source: Reprinted with permission of Boston Gear Company.

■ **FIGURE 3.25** A helical gear and integral shaft.

and disengagements cause spur gears to produce more drivetrain noise and vibration than other types of gears.

Helical gears are an alternative to spur gears; they offer the advantage of meshing more smoothly. Helical gears are similar to their spur counterparts in the sense that teeth are still formed on a cylinder, but in this case the teeth are not parallel to the gear's shaft. Instead, and as their name implies, the teeth on a helical gear are inclined at an angle so that each tooth wraps on the gear's cylinder in the shape of a shallow helix. A single helical gear is shown in Figure 3.25. With the same objective of having teeth mesh gradually, bevel gears (as in Figure 3.24) can likewise be formed with spiral instead of straight teeth.

Helix angle

- - - - - - - - - - - - - - - ➤

Spiral bevel gear

- - - - - - - - - - - - - - - ➤

Helical gears are evidently more complex to analyze and manufacture than spur gears. On the other hand, helical gearsets do have the advantage that they produce less noise and vibration in machinery. Automobile automatic transmissions, for instance, are typically constructed using both external and internal helical gears for precisely that reason. In a helical gearset, tooth-to-tooth contact starts at one edge of a tooth and

■ **FIGURE 3.26** A pair of crossed helical gears in mesh. The shafts for the two gears are perpendicular and offset from one another by the sum of the pinion's and gear's pitch radii.

■ **FIGURE 3.27**   A representative collection of helical and crossed helical gears.
Source: Reprinted with permission of Boston Gear Company.

proceeds gradually across its width, thus smoothing out the engagement and disengagement of teeth. Another attribute of helical gears is that they are capable of carrying greater torque and power when compared to similarly sized spur gears because the tooth-to-tooth contact forces are spread over more surface area.

Crossed helical gear
------------------→
We have seen how helix-shaped teeth can be formed on gears that are mounted on parallel shafts, but gears that are attached to perpendicular shafts can also incorporate helical teeth. The crossed helical gears of Figure 3.26 connect two perpendicular shafts, but unlike the bevel gearset application, the shafts here are offset from one another and extensions to their centerlines do not intersect. A sampler of helical gears is shown in Figure 3.27.

Worm Gearsets

If the helix angle on a pair of crossed helical gears is large enough, the resulting pair is called a *worm* and *worm gear*. Figure 3.28 illustrates this type of gearset where the worm itself has only one tooth that wraps several times around a cylindrical body,

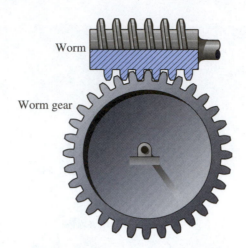

Worm

Worm gear

■ **FIGURE 3.28** A worm gearset.

■ **FIGURE 3.29** A worm having one continuous tooth.
Source: Reprinted with permission of Boston Gear Company.

similar to a thread of a screw. For each revolution of the worm (Figure 3.29), the worm gear (Figure 3.30) advances by just one tooth in its rotation.

Worm gearsets (Figure 3.31) can operate with large speed-reduction ratios. For instance, if the worm gear has 50 teeth, the speed reduction for a gearset as in Figure 3.28 would be 50-fold. The ability to package a geartrain with large speed reduction into a small physical space is an attractive feature of worm gearsets. However, the tooth profiles within worm gearsets are not involutes. Significant sliding occurs between

■ **FIGURE 3.30** Worm gears.
Source: Reprinted with permission of Boston Gear Company.

■ **FIGURE 3.31** A representative collection of worms and worm gears.
Source: Reprinted with permission of Boston Gear Company.

teeth during meshing, and that friction results in power loss, heating, and inefficiency compared to other types of gears.

Self-locking - - - - - - - - - - - - - - - →

It is also possible to design worm gearsets such that they can be driven in only one direction, namely, from the worm to the worm gear. For such a self-locking gearset, it would not be possible to reverse the power flow and cause the worm gear to backdrive the worm. This one-way drive capability is sometimes exploited in such applications as hoists or jacks where it is necessary to mechanically prevent the system from being backdriven. Not all worm gearsets are self-locking; this characteristic depends on such factors as the helix angle, the amount of friction between the worm and worm gear, and the presence of vibration.

3.5   BELT AND CHAIN DRIVES

Like gearsets, belt and chain drives are also available for transferring rotation, torque, and power between shafts. They are used in such applications as compressors, appliances, machine tools, automotive engines (Figure 3.32), and sheet metal rolling mills. Some attributes of belt and chain drives include the abilities to isolate different

■ **FIGURE 3.32**   Belt and chain drives are used on the front end of this 3.4-liter, twin dual overhead cam, 24-valve automobile engine.

Source: © 2001 General Motors Corporation. Used with permission of GM Media Archives.

elements in a drivetrain from shock, to have relatively long working distances between shaft centers, and to tolerate misalignment between adjacent shafts. These favorable characteristics stem largely from the belt's or chain's flexibility.

v-belt
----------------➤

Sheave
----------------➤

The common type of power transmission belt shown in Figure 3.33 is called a *v-belt*, named appropriately after the wedge-shaped appearance of its cross section. The grooved pulleys on which the v-belt rides are called *sheaves*. In order to have efficient transfer of power between shafts, the belt must be tensioned and have good frictional contact with its sheaves. In fact, the cross section is designed to wedge the belt down into the groove of the sheave and increase the friction force between the two. The capability of v-belts to transfer load between shafts is determined by the wedge angle, β, and the point at which the belt will begin to slip on a sheave, an event that is often accompanied by high-pitched squeal noise. The exterior of a v-belt is made from a synthetic rubber material to generate friction between the belt and sheaves. These belts are also internally reinforced with fiber or wire cords, which carry the bulk of the tension.

While v-belts are well suited for transmitting power, some amount of slippage invariably occurs between the belt and sheaves because the only contact between the two is friction. If you initially place small paint marks adjacent to one another on the belt and sheave, those marks will shift and separate from one another after some operation. Slippage between gears, on the other hand, cannot occur because of the direct

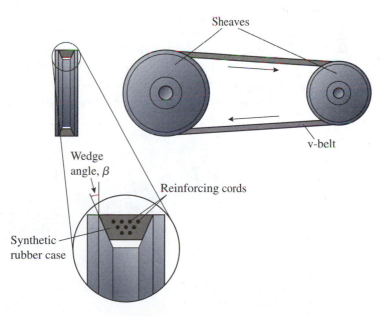

FIGURE 3.33   A v-belt and its sheaves.

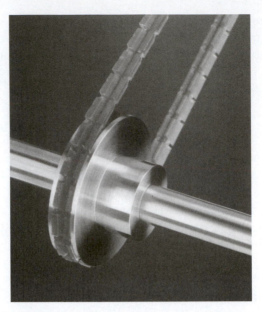

■ **FIGURE 3.34** A segmented v-belt and its sheave.
Source: Reprinted with permission of W. M. Berg, Incorporated.

Synchronous rotation --------->

mechanical engagement between their teeth. Geartrains are said to be a synchronous method of rotation, meaning that the input and output shafts rotate exactly together.

Belt slippage is not a problem if one is interested only in power transmission, as in a gasoline engine that drives a compressor or generator. On the other hand, for such precision applications as robotic manipulators and valve timing in automotive engines,

Timing belt --------->

shaft rotations must remain perfectly synchronized. To address that need, timing belts are formed with molded teeth that mesh into matching grooves on their sheaves. Timing belts, shown in Figure 3.35, combine some of the

■ **FIGURE 3.35** A magnified view of the teeth on a timing belt. The reinforcing cords are exposed at the belt's cross section.
Source: Reprinted with permission of W. M. Berg, Incorporated.

■ **FIGURE 3.36**   A chain and sprocket used in a power transmission drive.

best features of belts—mechanical isolation and long working distances between shaft centers—with a gearset's ability to provide synchronous motion.

Chain drive

Chain drives (Figure 3.36) can also be used when synchronous motion is required and, further, when high torque or power must be applied. Because of their metallic link construction, chains can generally carry greater forces than belts, and they are able to withstand high-temperature environments as well.

3.6   MACHINE TOOLS

In addition to being familiar with the hardware components described in the previous sections, to be an effective engineer you should also have some hands-on experience with the machine tools that one day will be used to fabricate the products that you design. In this section, we discuss drill presses, band saws, lathes, and mills, which are some of the tools available to perform standard machining operations in prototyping and production shops. Each of these tools is based on the principle of removing unwanted material from a work piece through the cutting action of sharpened bits or blades.

Mass production

The fabrication techniques that an engineer selects for a certain product will depend, in part, on the time and expense of setting up the tooling. Some devices (for instance, air-conditioner compressors, microprocessors, hydraulic valves, and tires) are mass-produced, implying a process that is based on widespread mechanical automation (Figure 3.37). Mass production might be what you first imagine to take place in a large factory, and the manufacture of automobile engines is a prime example. The assembly line in that case comprises custom tools, fixtures, and specialized processes that are capable in the end of producing only certain types of engines for certain vehicles. The assembly line will be set up so that only a small number of operations need to be performed in any one work area before the engine moves on to the next stage. Because finished engines will be produced at a relatively high rate, it is cost-effective for a company to allocate a large amount of factory floor space and many expensive machine tools to the production line, each of which might only drill a few holes or make a few welds. Aside from hardware

■ **FIGURE 3.37**   Robots automate welding on a mass production line for automobile frames.
Source: © 2001 General Motors Corporation. Used with permission of GM Media Archives.

that is produced through mass manufacturing, other products are either unique (for instance, the Hubble space telescope), or made in relatively small quantities (such as commercial jetliners). The best production method for a given product will ultimately depend on the quantity to be produced, the allowable cost, and the level of part-to-part variability that is acceptable.

Drill press
- - - - - - - - - - - - - - - - ▸

           The drill press shown in Figure 3.38 bores holes into a work piece. A drill bit is held in the rotating chuck, and as a machinist turns the pilot wheel, the bit is lowered into the surface of the work piece. The bit removes material in small chips and creates a hole such as those shown in the cross sections of Figure 3.39. As should be the case whenever metal is machined, the point where the bit cuts into the work piece is lubricated; the oil reduces friction, and it also helps to remove heat from the cutting region. For safety reasons, vises and clamps securely hold the work piece as the hole is bored in order to prevent material from shifting unexpectedly.

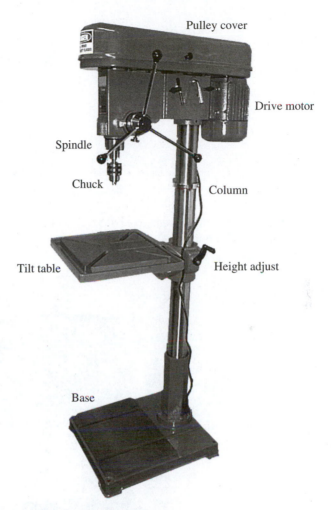

Pulley cover

Drive motor

Spindle

Chuck

Column

Tilt table

Height adjust

Base

■ **FIGURE 3.38** Major elements of a drill press.

Band saw

Lathe

A band saw makes rough cuts through metal or plastic stock. The configuration of a typical industrial band saw is shown in Figure 3.40. The blade is a long, continuous loop having sharp teeth on one edge, and it rides on the drive and idler wheels. A variable-speed motor enables the operator to adjust the blade's speed depending on the type and thickness of material that will be cut. A tilting table is used for support. The machinist feeds the work piece into the blade, and guides it by hand, to make straight or slightly curved cuts. When the blade becomes dull and needs to be replaced, or if it should break, the band saw's internal blade grinder and welder are used to clean up the blade's ends, connect them, and form a loop.

A machinist's lathe holds a work piece and rotates it as the sharpened edge of a tool cuts and removes material. Some applications of a lathe include the production of shafts, the turning and resurfacing of disk brake rotors, and

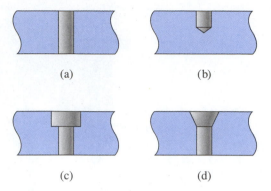

■ FIGURE 3.39 (a) Through, (b) blind, (c) counterbored, and (d) countersunk holes can be produced with a drill press.

■ FIGURE 3.40 Major elements of a band saw.

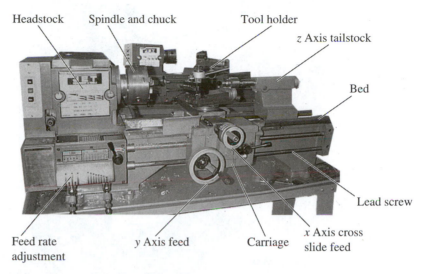

Headstock   Spindle and chuck   Tool holder   z Axis tailstock   Bed   Lead screw   x Axis cross slide feed   Carriage   y Axis feed   Feed rate adjustment

■ **FIGURE 3.41**   Major elements of a machinist's lathe.

thread cutting. A lathe with typical features is shown in Figure 3.41. The headstock and chuck hold one end of a piece of solid round bar stock, for instance, and the drive mechanism spins it rapidly about its centerline. The tailstock can be used to provide support to the otherwise free end of a long work piece. As a cutting tool is fed against the bar stock and moved along its length, the diameter of the work piece is reduced to a desired dimension. Shoulders that will locate bearings on a shaft, grooves for holding retaining clips, and sharp changes in diameter of a stepped shaft can each be made in this manner.

Mill
- - - - - - - - - - - - - - ►

A mill is useful for machining the rough surfaces of a work piece flat and smooth and for shaping them with slots, grooves, and holes. Shown in Figure 3.42, the mill is a versatile machine tool. The work piece is held by a vise on an adjustable table so that the part can be accurately moved in three directions (in the plane of the table and perpendicular to it) to locate the work piece beneath the cutting bit. In a typical application, a piece of metal plate would be cut to approximate shape with a bandsaw, and the mill would be used to machine the surfaces and edges smooth, square, and to the final dimensions.

Lead screw
- - - - - - - - - - - - - - ►

A machine component known as a lead screw is often used in the mill's internal mechanisms for positioning the work table beneath the spindle. One such application is shown in Figure 3.43. The lead screw converts rotation of a shaft, produced either by a hand wheel or an electric motor, into the straight-line motion of the work table. Figure 3.44 depicts several lead screws. As the lead screw turns and engages the nut (which is not allowed to rotate), the nut moves along the screw with each rotation by an amount equal to the distance between threads.

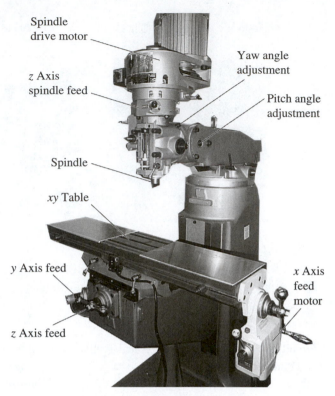

■ FIGURE 3.42 Major elements of a milling machine.

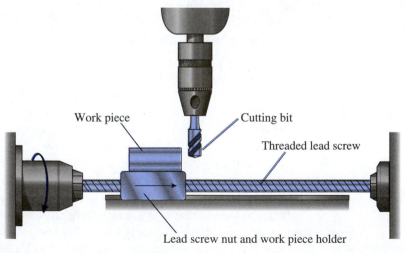

■ FIGURE 3.43 A lead screw converts a motor's rotation into the straight-line motion of the table and work piece in a mill.

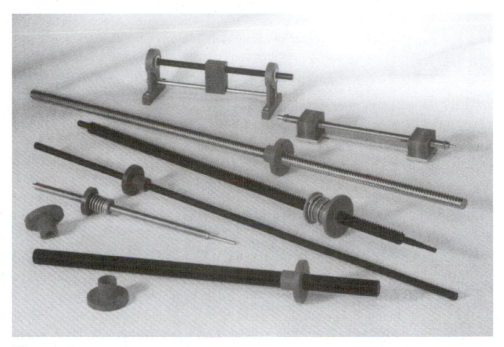

■ **FIGURE 3.44**   A collection of different lead screws.
Source: Reprinted with permission of W. M. Berg, Incorporated.

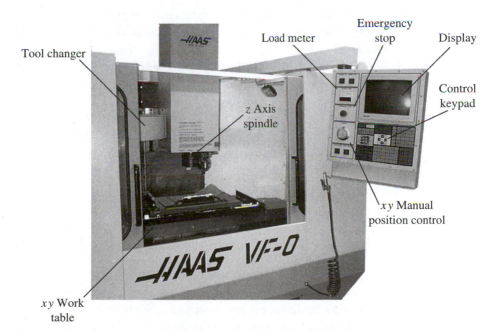

■ **FIGURE 3.45**   A computer-controlled mill can produce hardware directly from the machining instructions created by certain computer-aided engineering software packages.

By using electric motors to drive lead screws and position the work piece or cutting bit, mills and other machine tools can be controlled by computers. In that manner, a shop operation can be automated to achieve high precision or to complete a repetitive task on a large number of parts. Figure 3.45 (on the previous page) shows an example of such a computer numerically controlled (CNC) mill. A CNC mill performs the same types of operations as a conventional mill, but instead of being manually operated, it is programmed either through entry on a keypad or by downloading machining instructions that were created by computer-aided engineering software. CNC machine tools offer the potential to seamlessly produce physical hardware directly from a computer-generated drawing. With the ability to quickly reprogram machine tools, even a small general-purpose shop can produce a variety of high-quality machine components. In Chapter 8, we explore a case study in which a suite of computer-aided engineering software packages was used to design, analyze, and ultimately manufacture a mechanical component.

CNC mill
- - - - - - - - - - - - - ->

SUMMARY

In this chapter, we have described a sampler of the machine components and tools that arise in mechanical engineering. Although not exhaustive, these summaries of bearings, gears, and power transmission components on the one hand—and of shop tools on the other—should help you understand and better describe to others the anatomy and function of common machines. Bearings, gears, shafts, and the like each have special properties and terminology, and mechanical engineers need to be fluent with such building blocks in order to select the component that is best suited to a particular application.

In Chapters 4–7, we will apply physical laws and the tools of engineering science (namely, the numbers and equations side of engineering) to analyze engineering problems that are based in part on some of the components and tools described in this chapter. Later, in Chapter 8, we return to the subject of mechanical design and investigate the major steps involved in defining a design problem, brainstorming potential solutions, developing a detailed design, and protecting the new technology through patents.

SELF-STUDY AND REVIEW

1. Explain the differences between ball, straight roller, tapered roller, and thrust bearings. Give examples of real situations in which you would select one to be used instead of another.
2. What is the purpose of a bearing's separator?
3. Make a cross-sectional drawing of a tapered roller bearing.
4. Why are flexible couplings sometimes used when connecting shafts in a drivetrain?
5. Make an accurate sketch for the shape of a spur gear's teeth.
6. What is the fundamental property of gearsets?
7. Define the terms *diametral pitch* and *module*.

8. What are a rack and pinion?
9. What are the characteristics of helical gears, and how do they differ from spur gears?
10. Make a sketch to show the difference in shaft orientations when bevel gears and crossed helical gears are used.
11. Under what circumstances would you select a v-belt to be used instead of a timing belt in a machine and vice versa?
12. Briefly describe the operation of a drill press, band saw, lathe, and mill.
13. What is a lead screw?

4 Forces in Structures and Fluids

4.1 OVERVIEW

Mechanical engineers use mathematics and physical laws to design hardware better and faster than would be possible otherwise. By applying the principle of force balance, for instance, an engineer can often analyze a design to a reasonable level of accuracy before any hardware is built. Engineers reduce the time and expense associated with constructing and testing prototypes by first refining their designs on paper. Computer-aided engineering tools further increase the level of sophistication that is available for such analyses. With that perspective in mind, in this chapter we examine forces in structures and fluids as the second element of mechanical engineering.

This chapter introduces you to the subject of mechanics, a topic that encompasses forces which act on structures and machines and their tendency either to remain stationary or move. We will explore the properties of forces and the problem-solving skills that are needed to understand their effects on engineering hardware. After developing the concepts of force systems, moments, and static equilibrium, we will calculate the magnitudes and directions of forces acting on, and within, simple structures and machines. In Section 4.5, we discuss the important forces known as buoyancy, drag, and lift that arise when fluids (such as water or air) interact with a structure.

In short, the process of analyzing forces is a first step taken by engineers when deciding whether a certain piece of hardware will operate reliably, or break, and whether or how a machine or vehicle will move (Figure 4.1). After completing this chapter, you should be able to:

- Describe a force in terms of its rectangular and polar components.
- Calculate the resultant of a system of forces by using the vector algebra and polygon methods.
- Calculate the moment of a force about a point using the perpendicular lever arm and moment component methods.
- Understand the requirements for equilibrium, and calculate unknown forces.
- Understand the physical meaning of the density and viscosity properties of a fluid.

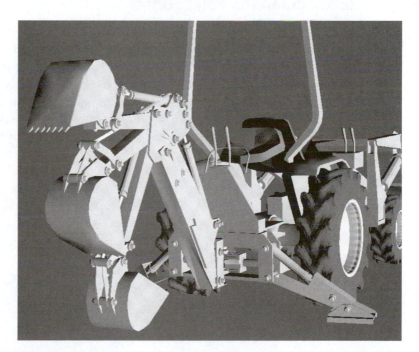

■ **FIGURE 4.1**   Heavy construction equipment is designed to support the large forces developed during operation.

Source: Reprinted with permission of Mechanical Dynamics, Incorporated, and by Caterpillar, Incorporated.

- Explain the difference between laminar and turbulent flowing fluids.
- Calculate and describe the dimensionless Reynolds and Mach numbers.
- Discuss the fluid forces known as buoyancy, drag, and lift, and calculate them in certain applications.

4.2 FORCES AND RESULTANTS

In this section, we discuss properties of forces, different ways of representing them, and two methods for calculating the net effect of several forces that act together. Forces are vectors because their physical action involves both magnitude and direction. You probably already have a good intuition for the tendency of a force to pull or push an object in a certain direction. The magnitude of a force is measured by pounds (lb) or ounces (oz) in the USCS, and by newtons (N) in the SI. Conversion factors between the two systems of units are summarized in Table 4.1. Reading off the table's first line, for instance, we see that 1 lb = 16 oz = 4.448 N, and from the third line that 1 N = 3.597 oz = 0.2248 lb. In this chapter and the following ones, we will use conversion tables of this format to relate various engineering quantities in the USCS and SI.

Rectangular and Polar Forms

Vector notation
- - - - - - - - - - - - - - ➤

Rectangular components
- - - - - - - - - - - - - - ➤

In this textbook, we will use boldface notation—**F**—to denote force vectors. A common method to describe a force is in terms of its horizontal and vertical components. These are denoted by F_x and F_y in Figure 4.2. Once we set the directions for the x and y axes, the force **F** can be broken down into its components along those directions. The projection of **F** in the horizontal direction (the x axis) is called F_x, and the vertical projection (y axis) is called F_y. When you assign numerical values to F_x and F_y, you have fully described the force **F**. In fact, the pair of numbers (F_x, F_y) is just the coordinates of the force vector's tip.

Unit vectors
- - - - - - - - - - - - - - ➤

In Figure 4.2, the unit vectors **i** and **j** are used to indicate the directions in which F_x and F_y act. Vector **i** points along the positive x direction, and **j** is a vector pointing in the positive y direction. Just as F_x and F_y provide

TABLE 4.1 Conversion Factors Between USCS and SI Units for Force

lb	oz	N
1	16	4.448
0.0625	1	0.2780
0.2248	3.597	1

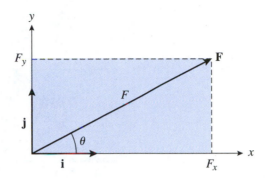

FIGURE 4.2 Representing a force vector in terms of its rectangular components (F_x, F_y), and its polar components (F, θ).

information about the magnitudes of the horizontal and vertical components, the unit vectors give information about the directions of those components. The unit vectors are so named because they have a length of 1. By combining components and unit vectors, the force **F** is described completely using vector algebra notation as

$$\mathbf{F} = F_x\mathbf{i} + F_y\mathbf{j} \tag{4.1}$$

Polar components

Vector magnitude

Vector direction

In an alternative view, rather than thinking about a force in terms of how hard it pulls rightward (F_x) and upward (F_y), you could explain how hard it pulls and in which direction. The latter viewpoint is based on polar coordinates. As also shown in Figure 4.2, **F** acts at the angle θ, which is measured relative to the horizontal axis. The magnitude or length of the force vector is a scalar quantity, and it is denoted by $F = |\mathbf{F}|$, where the $| \bullet |$ notation designates the vector's absolute value. By our convention, we write the (scalar) magnitude F in a plain typeface. Instead of specifying F_x and F_y, we can now view the force vector **F** in terms of the quantities F and θ. This representation is called the *polar component* (or magnitude-direction) form of the vector.

Referring to Figure 4.2, we relate the force's magnitude and direction to its horizontal and vertical components through the trigonometric equations

$$F_x = F\cos\theta \qquad \text{and} \qquad F_y = F\sin\theta \tag{4.2}$$

If we happen to know the force's magnitude and direction, these equations are used to determine its horizontal and vertical components. On the other hand, when we are given F_x and F_y, the magnitude and direction are calculated through the relationships

$$F = \sqrt{F_x^2 + F_y^2} \qquad \text{and} \qquad \theta = \tan^{-1}\left(\frac{F_y}{F_x}\right) \tag{4.3}$$

The inverse tangent operation in Equation (4.3) calculates the principal value and returns an angle between $-90°$ and $+90°$, namely, in either the first or fourth quadrants of the xy plane. Of course, a force could be oriented in any of the four quadrants. In problem solving, you will need to examine the positive or negative signs of F_x and F_y

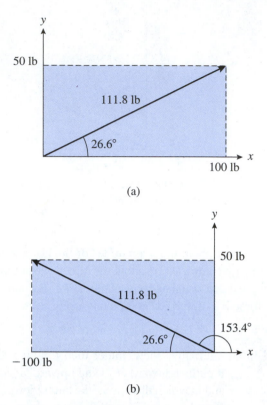

(a)

(b)

FIGURE 4.3   Determining the angle of action for a force that (a) lies in the first quadrant and (b) lies in the second quadrant.

and use them to determine the correct quadrant for θ. For instance, in Figure 4.3(a), $F_x = 100$ lb and $F_y = 50$ lb, and the force's angle of action is calculated as being $\theta = \tan^{-1}(0.5) = 26.6°$. That numerical value correctly lies in the first quadrant because both F_x and F_y are positive. On the other hand, in Figure 4.3(b), when $F_x = -100$ lb and $F_y = 50$ lb, you might be tempted to report an angle of $\tan^{-1}(-0.5) = -26.6°$. That value falls in the fourth quadrant, and it is incorrect as a measure of the force's orientation relative to the positive x axis. However, as is evident from Figure 4.3(b), **F** does form an angle of 26.6° relative to the negative x axis. The correct value for the force's angle of action relative to the positive x axis is $\theta = 180° - 26.6° = 153.4°$.

Resultants

Force system

A force system is a collection of several forces that simultaneously act on an object. Each force is combined with the others in order to describe their net effect, and the resultant **R** measures that overall or cumulative action. As an example, consider the mounting post and bracket of Figure 4.4. The three forces

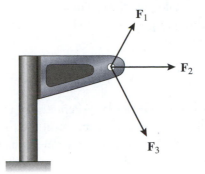

■ FIGURE 4.4 A mounting post and bracket that are loaded by three forces.

$\mathbf{F}_1$, $\mathbf{F}_2$, and $\mathbf{F}_3$ act in different directions and have different magnitudes. To determine whether or not the post is capable of supporting those forces, an engineer first needs to determine their net effect. With N individual forces denoted by $\mathbf{F}_i$ $(i = 1, 2, \ldots, N)$, they are summed according to

Resultant

$$\mathbf{R} = \mathbf{F}_1 + \mathbf{F}_2 + \cdots + \mathbf{F}_N$$

$$= \sum_{i=1}^{N} \mathbf{F}_i \tag{4.4}$$

by using the rules of vector algebra. The summation can be carried out by the vector algebra or vector polygon approaches, as described next. In different problem-solving situations, you will find that one approach might be simpler than another, and the use of more than one approach is often a good idea for double-checking your calculations.

Vector Algebra Method   In this technique, each force $\mathbf{F}_i$ is broken down into its horizontal and vertical components, which we label as F_{xi} and F_{yi} for the ith force. The resultant's horizontal component R_x is found by

$$R_x = \sum_{i=1}^{N} F_{xi} \tag{4.5}$$

Likewise, we separately sum the vertical components through

$$R_y = \sum_{i=1}^{N} F_{yi} \tag{4.6}$$

The resultant force is then expressed as $\mathbf{R} = R_x\mathbf{i} + R_y\mathbf{j}$. If we are interested in the magnitude R and direction θ of $\mathbf{R}$, we apply the expressions

$$R = \sqrt{R_x^2 + R_y^2} \quad \text{and} \quad \theta = \tan^{-1}\left(\frac{R_y}{R_x}\right) \tag{4.7}$$

which are analogous to Equation 4.3. As before, the actual value for θ is found after considering the positive and negative signs of R_x and R_y so that θ is placed in the correct quadrant.

Vector Polygon Method Alternatively, the resultant of a force system can be found by sketching a polygon to represent addition of the $\mathbf{F}_i$ vectors. The magnitude and direction of the resultant are then determined by applying rules of trigonometry to the polygon's geometry. Referring to the mounting post of Figure 4.4, the vector polygon

Head-to-tail rule

for those three forces is drawn by adding the individual $\mathbf{F}_i$'s in a chain according to the head-to-tail rule. In Figure 4.5, the starting point is labeled on the page, the three forces are added in turn, and the endpoint is also labeled on the page. The order in which the forces are added to the diagram does not matter insofar as the final result is concerned, but your diagrams will appear visually different for various addition sequences. The endpoint of the diagram is located at the tip of the last vector that was added to the chain. As indicated in Figure 4.5, the resultant $\mathbf{R}$ extends from the start of the chain to its end. The action of $\mathbf{R}$ on the bracket is equivalent to the combined effect of the three forces. The magnitude and direction of the resultant are determined by applying trigonometric identities to the polygon's shape. Some of the relevant equations for right and oblique triangles are reviewed in Appendix B.

We can often obtain reasonably accurate results by summing vectors on a drawing that is made to scale, for instance, 1 inch on the drawing corresponds to 100 pounds of force. Such drafting tools as a protractor, scale, and straightedge should be used to construct the polygon and to measure the magnitudes and directions of unknown quantities. It is certainly acceptable to use a purely graphical approach when solving engineering problems, provided that you can determine the answer to a fair number of significant digits.

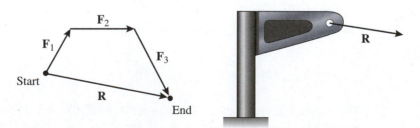

FIGURE 4.5 The resultant $\mathbf{R}$ extends from the start to the end of the chain formed by adding $\mathbf{F}_1$, $\mathbf{F}_2$, and $\mathbf{F}_3$ together.

EXAMPLE 4.1

The eyebolt shown in Figure 4.6 is bolted to a thick base plate, and it supports three steel cables with tensions 150 lb, 350 lb, and 400 lb in the directions shown. Determine the resultant force that acts on the eyebolt by using the (a) vector algebra and (b) vector polygon solution approaches. The unit vectors **i** and **j** are oriented as shown.

SOLUTION

(a) In the vector algebra approach, we first break the forces down into their components along the **i** and **j** directions:

$$\mathbf{F}_1 = (400 \cos 45°)\mathbf{i} + (400 \sin 45°)\mathbf{j} \text{ lb}$$

$$= 283\mathbf{i} + 283\mathbf{j} \text{ lb}$$

$$\mathbf{F}_2 = -(350 \sin 20°)\mathbf{i} + (350 \cos 20°)\mathbf{j} \text{ lb}$$

$$= -120\mathbf{i} + 329\mathbf{j} \text{ lb}$$

$$\mathbf{F}_3 = -150\mathbf{i} \text{ lb}$$

By adding the respective components, the resultant becomes

$$\mathbf{R} = (283 - 120 - 150)\mathbf{i} + (283 + 329)\mathbf{j} \text{ lb}$$

or $\mathbf{R} = 13\mathbf{i} + 612\mathbf{j}$ lb. The magnitude is $R = \sqrt{13^2 + 612^2}$ lb $= 612$ lb. The force acts at angle $\tan^{-1}(612/13) = 88.78°$, measured counterclockwise from the **i** direction.

(b) In the second solution approach, a polygon is sketched and the known magnitudes and directions of the forces are labeled on the diagram. The resultant is found

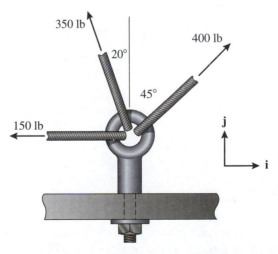

FIGURE 4.6 An eyebolt that is loaded by three tension cables in Example 4.1.

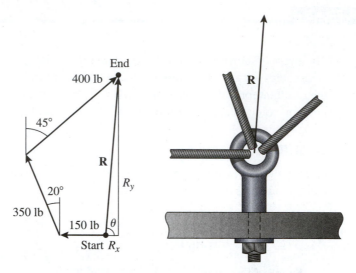

■ **FIGURE 4.7** The vector polygon used in calculating the resultant force's magnitude and direction in Example 4.1.

by applying the laws of right triangles. This type of solution therefore includes both a drawing and a calculation. Figure 4.7 shows the polygon for the situation at hand. The known side lengths (force magnitudes) and angles (force directions) are written on the diagram after the vectors have been drawn using the head-to-tail rule. The components R_x and R_y are found by analyzing the polygon's geometry. The equations for sums of the components become (with each term having the units of lb)

$$-150 - 350 \sin 20° + 400 \cos 45° - R_x = 0$$

in the horizontal direction, and

$$350 \cos 20° + 400 \sin 45° - R_y = 0$$

in the vertical direction. Here we must be careful to indicate the correct signs for each term. Positive values refer to a component directed rightward or upward, and negative values refer to a component directed leftward or downward. From these relations, the resultant's components are again found to be $R_x = 13$ lb and $R_y = 612$ lb. ■

EXAMPLE 4.2

The control lever rotates when the 10-lb and 25-lb forces act on it as shown in Figure 4.8(a). Determine the magnitude and direction of the resultant by using the vector polygon approach.

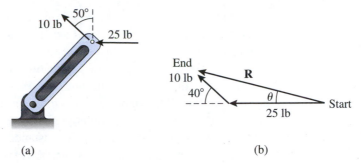

(a)   (b)

■ **FIGURE 4.8** Vector polygon used in calculating the resultant force's magnitude and direction in Example 4.2.

SOLUTION

The two forces are combined in a sketch using the head-to-tail rule. In Figure 4.8(b), the 25-lb force is sketched first and the 10-lb force is added at the angle of 50° from vertical. The resultant **R** extends from the tail of the 25-lb force vector (which is labeled as the start point) to the head of the 10-lb force vector (labeled as the endpoint). The three vectors form a side–angle–side triangle. Applying the law of cosines from Appendix B, we solve for the side length R through

$$R^2 = (10\ \mathrm{lb})^2 + (25\ \mathrm{lb})^2 + 2(10\ \mathrm{lb})(25\ \mathrm{lb})\cos 40°$$

from which $R = 33.29$ lb. The angle θ at which **R** acts is similarly determined by applying the law of sines

$$\frac{\sin(180° - 40°)}{33.29\ \mathrm{lb}} = \frac{\sin\theta}{10\ \mathrm{lb}}$$

The resultant acts at the angle $\theta = 11.13°$ shown in Figure 4.8(b). ■

4.3 MOMENT OF A FORCE

When you are trying to loosen a frozen bolt, the bolt is more easily turned when a wrench with a long handle is used. The tendency of a force (in this case, applied to the handle's end) to make an object rotate is called a *moment*. The magnitude of a moment depends both on the force that is applied and on the lever arm that separates the force from a pivot point.

Perpendicular Lever Arm

The magnitude of a moment is found from its definition

$$M_o = Fd \tag{4.8}$$

Moment magnitude
- - - - - - - - - - - - ▸

where M_o is the moment of the force about point O, F is the magnitude of the force, and d is the perpendicular lever arm distance from the force's line

Lever arm

Torque

Moment units

of action to point O. The term *torque* can also be used to describe the effect of a force acting over a lever arm, but mechanical engineers generally reserve *torque* to describe moments that cause rotation of a shaft in a motor, engine, or gearbox. We will discuss those applications in Chapter 7.

Based on Equation 4.8, the unit for M_o is the product of force and distance. In the USCS, the unit for a moment is in·lb or ft·lb. In the SI, the unit N·m is used, and various prefixes are applied when the numerical value is either very large or very small. For instance, 5000 N·m = 5 kN·m and 0.002 N·m = 2 mN·m. Conversion factors between the two systems of units are shown in Table 4.2, where you can see that 1 ft·lb = 1.356 N·m. Work and energy, which are other quantities that arise in mechanical engineering, also have units that are the product of force and distance. For instance, when working in the SI, a joule (J) is defined as one newton-meter. It is the amount of work performed by a 1 newton force that moves through a distance of 1 meter. However, the physical quantities of work and energy are quite different from moments and torques, and in order to be clear when distinguishing them, the unit of N·m (and not J) should be used in the SI for moment and torque.

The expression $M_o = Fd$ can best be understood by applying it to a specific structure. In Figure 4.9(a), the force **F** is directed generally downward and to the right on the bracket. One might be interested in the moment of **F** about the base of the support post, which is labeled in the figure as point O. The structure could break at that location, and an engineer would examine the post to make sure that it can support **F**. The moment is calculated based upon both the magnitude of **F** and the perpendicular offset distance d between the force's line of action and point O. In fact, **F** could be applied to the bracket at any point along its line of action, and the moment produced about O would remain unchanged because d would likewise not change. The direction of the moment is clockwise because **F** tends to cause the post to rotate that way (even though the rigid mounting would prevent the post from actually moving in this case).

Line of action

In Figure 4.9(b) the direction of **F** has been changed. The force's line of action now passes directly through point O, and the offset distance becomes $d = 0$. No moment is produced, and the force tends to pull the post directly out of its base. In

TABLE 4.2 Conversion Factors Between USCS and SI Units for Moment or Torque

in·lb	ft·lb	N·m
1	0.0833	0.1130
12	1	1.356
8.851	0.7376	1

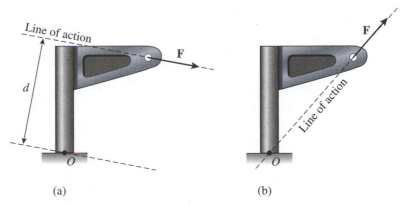

■ **FIGURE 4.9**   Calculating the moment of force **F** about point 0. (a) The line of action of **F** is separated from 0 by the perpendicular lever arm distance d. (b) The line of action of **F** passes through 0, and $M_0 = 0$.

short, the orientation of a force as well as its magnitude must be taken into account when calculating a moment.

EXAMPLE 4.3

The open-end wrench shown in Figure 4.10 tightens a hexagonal head nut and bolt. Calculate the moments produced by the 35-lb force about the center of the nut when the force is applied to the wrench in the two orientations shown. The overall length of the handle, which is inclined slightly upward, is $6\frac{1}{16}$ in. long between centers of the open and closed ends.

SOLUTION

(a) In Figure 4.10(a), the 35-lb force acts vertically downward. The perpendicular distance from the center of the nut to the force's line of action is $d = 6$ in. The incline and length of the wrench's handle are immaterial insofar as calculating d is concerned; the handle's length is not necessarily the same as the perpendicular lever arm distance. The moment has magnitude

$$(35 \text{ lb})(6 \text{ in.}) = 210 \text{ in} \cdot \text{lb} = 17.5 \text{ ft} \cdot \text{lb}$$

which is directed clockwise (CW).

(b) In Figure 4.10(b), the force has shifted to an inclined angle, and its line of action has changed so that d is measured to be $5\frac{3}{8}$ in. The moment is reduced to

$$(35 \text{ lb})(5.375 \text{ in.}) = 188 \text{ in} \cdot \text{lb} = 15.7 \text{ ft} \cdot \text{lb}$$

We report the answer as $M_0 = 15.7$ ft·lb (CW).

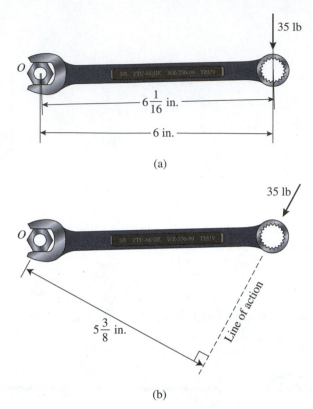

FIGURE 4.10 Calculating the perpendicular lever arm distances for forces acting on a wrench in Example 4.3.

Moment Components

Just as we can break a force into rectangular components, it is sometimes useful to calculate a moment as the sum of its components. The moment is determined as the sum of portions that are associated with the two components of the force, rather than the full resultant value of the force. A motivation for calculating the moment in this manner is that it is often easier to find the lever arms for individual components than for the resultant force. When applying this technique, it is necessary to use a sign convention and keep track of whether the contribution made by each force component is clockwise or counterclockwise.

To illustrate this method, we calculate the moments about point O of the forces shown in Figure 4.11. We first choose the following sign convention: A moment that is directed clockwise is positive, and a counterclockwise moment is negative.

Sign convention

The sign convention is just a bookkeeping tool for combining the various clockwise and counterclockwise moment components. Any contribution to the moment about O that acts clockwise is given a positive sign, and any contribution

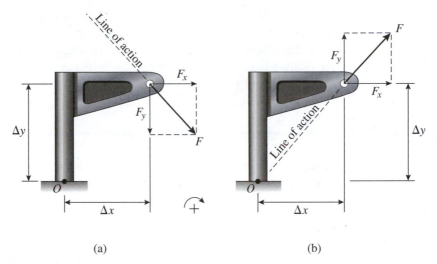

(a)   (b)

■ **FIGURE 4.11**   Calculating moments based on components. (a) Both F_x and F_y create clockwise moments about point O. (b) F_x exerts a clockwise moment, but F_y exerts a counterclockwise moment.

in the other direction is negative. This choice of positive and negative directions is arbitrary, and we could just as easily have selected the counterclockwise direction as being positive. However, once the sign convention is chosen, we stick with it and apply it consistently.

In Figure 4.11(a), the force is broken down into the components F_x and F_y. Rather than determine the distance from point O to the line of action of **F**, which might involve a geometrical construction that we want to avoid, we instead calculate the individual lever arm distances for F_x and F_y, which are more straightforward. Keeping track of the sign convention, the moment about O becomes $M_o = F_x \Delta y + F_y \Delta x$. Each contribution to M_o is positive because F_x and F_y each tend to cause clockwise rotation. Their effects combine constructively.

The orientation of **F** has been changed in Figure 4.11(b). While the component F_x continues to exert a positive moment, F_y now tends to cause counterclockwise rotation about O. It therefore makes a negative contribution, and the net moment becomes $M_o = F_x \Delta y - F_y \Delta x$. Here the two components combine in a deconstructive manner. For the particular orientation in which $\Delta y / \Delta x = F_y / F_x$, the two terms precisely cancel. The moment in that case is zero because the line of action for **F** passes directly through O, as in Figure 4.9(b). In the general case of the moment components method, we write

$$M_o = \pm F_x \Delta y \pm F_y \Delta x \tag{4.9}$$

and the positive and negative signs are selected depending on whether the component tends to cause clockwise or counterclockwise rotation.

Regardless of which method you use to calculate a moment, when reporting an answer you should state (1) the numerical magnitude of the moment, (2) the units, and (3) the direction. You can indicate the direction by using ± notation provided that you have also shown the sign convention on your drawing. The notation CW or CCW is also useful for denoting whether the moment acts clockwise or counterclockwise.

Determine the moment about the center of the nut as the 250-N force is applied to the handle of the adjustable wrench in Figure 4.12. Use (a) the perpendicular lever arm method, and (b) the moment components method.

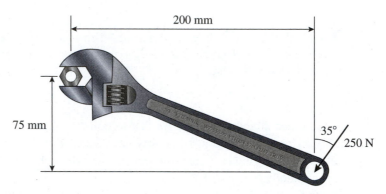

FIGURE 4.12 An adjustable wrench that is loaded by a 250-N force in Example 4.4.

SOLUTION

(a) As shown in Figure 4.13(a), we denote the center of the nut as point A, and the point of application of the force as point B. Using the dimensions given, distance AB is calculated as $\sqrt{75^2 + 200^2}$ mm = 214 mm. Although this is the distance from A to the location at which the force is applied, it is not the perpendicular lever arm distance d. For this reason, we need to calculate the length of segment AC. Because the applied force is tilted 35° from vertical, a line perpendicular to the force is oriented 35° from horizontal. As depicted in Figure 4.13(a), line AB lies at the angle $\tan^{-1}(75/200) = 20.6°$ below horizontal, and so it is offset by $35° - 20.6° = 14.4°$ from AC. The correct lever arm distance therefore becomes $d = (214 \text{ mm})\cos 14.4° = 207$ mm. The wrench's moment is $M_A = (250 \text{ N})(0.207 \text{ m}) = 51.8$ N·m, directed clockwise (CW).

(b) In Figure 4.13(b), the 250-N force is broken into its components. The horizontal portion is $(250 \text{ N})\sin 35° = 143$ N, and the vertical component is $(250 \text{ N})\cos 35° = 205$ N. Those components are oriented leftward and downward, respectively, and they each exert clockwise moment contributions about A. In Figure 4.13(b), we have shown

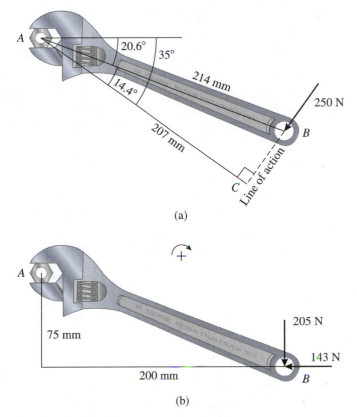

FIGURE 4.13   Calculating moments using the (a) perpendicular lever arm method and (b) moment components method in Example 4.4.

our positive sign convention for moments as being clockwise. By summing the moment of each component, we have

$$M_A = (143 \text{ N})(0.075 \text{ m}) + (205 \text{ N})(0.2 \text{ m}) = 51.8 \text{ N·m}$$

Because the net result is positive, the moment is directed clockwise.   ■

4.4   EQUILIBRIUM OF FORCES AND MOMENTS

With groundwork for the properties of forces and moments now in place, we next turn to the task of calculating (unknown) forces that act on structures and machines in response to other (known) forces that are present. This process involves applying the principle of static equilibrium to systems that are either stationary or moving at constant velocity. In either case, no acceleration is present, and in accordance with the laws of motion, the resultant force on the system is zero.

Particles and Rigid Bodies

A mechanical system can include either a single object (for instance, an engine piston) or multiple objects that are connected together (the entire engine). When the physical dimensions of the object are unimportant in calculating forces, the object is called a *particle*. This concept idealizes an object as being concentrated at a single point, rather than distributed over an extended area or volume. For the purposes of problem solving, a particle can therefore be treated as having negligible dimensions. On the other hand, if the length, width, and breadth of an object are important for the problem at hand, it is called a *rigid body*. When looking at the motion of the Space Shuttle as it orbits the Earth, for instance, the spacecraft is regarded as a particle because its dimensions are small compared to the size of the orbit. However, when the Shuttle is landing and engineers are interested in its aerodynamics and flight characteristics, the vehicle would instead be analyzed as a rigid body. Figure 4.14 illustrates the distinction between forces that are applied to a particle and to a rigid body; you can see how a force imbalance could cause the rigid body to rotate.

Force balance
- - - - - - - - - - - - - - - - →   A particle is in equilibrium if the forces acting on it balance with zero resultant. Because forces combine as vectors, the resultant must be zero in two perpendicular directions, which we label x and y:

$$\sum_i F_{xi} = 0 \qquad \text{and} \qquad \sum_i F_{yi} = 0 \tag{4.10}$$

Moment balance
- - - - - - - - - - - - - - - - →   For a rigid body to be in equilibrium, it is necessary that (1) the resultant of all forces is zero, and (2) the net moment is also zero. When those conditions are met, there is no tendency for the object either to move (in response to the forces) or to rotate (in response to the moments). The requirements for equilibrium of a rigid body involve Equation 4.10 and

$$\sum_i M_{oi} = 0 \tag{4.11}$$

The notation M_{oi} is used to denote the ith moment that is applied about point O. In summary, the two relationships in Equation 4.10 specify that no net force acts in

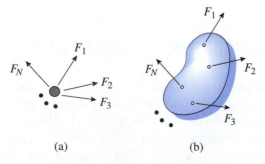

■ FIGURE 4.14   A schematic of *N* forces acting on a (a) particle and (b) rigid body.

either the x or y directions, and Equation 4.11 states that no net moment tends to cause rotation.

Sign conventions are a bookkeeping method to distinguish forces acting in opposing directions and moments that are oriented clockwise or counterclockwise. The summations in the equilibrium equations are conducted over all forces and moments present, whether or not their directions and magnitudes are known in advance. Forces that are unknown at the start of the problem are always included in the summation, algebraic variables are assigned to them, and the equilibrium equations are then applied to determine numerical values.

Mathematically speaking, the equilibrium equations for a rigid body comprise a system of three linear equations involving the unknown forces and moments. One implication of this characteristic is that it is possible to determine at most three unknown quantities when Equations 4.10–4.11 are applied to a single rigid body. By contrast, when applying the equilibrium requirements to a particle, the moment equation

Independent equations
- - - - - - - - - - - - - - ➤

is not used. Therefore, only two independent equations are produced, and only two unknowns can be determined. It is not possible to obtain more independent equations of equilibrium by resolving moments about an alternative point or by summing forces in different directions. The additional equations will still be valid, but they will simply be combinations of the other (already derived) ones. As such, they will provide no new information. When you solve equilibrium problems, you should always check to be sure that you do not have more unknown quantities than independent equations.

Free Body Diagrams

FBD
- - - - - - - - - - - - - - ➤

Free body diagrams (abbreviated FBD) are sketches used to analyze the forces and moments that act on structures and machines, and their construction is an important skill. The FBD is used to identify the mechanical system that is being examined and to represent all of the known and unknown forces that are present. Three main steps are followed when a FBD is drawn:

1. Select an object that will be analyzed by using the equilibrium equations. Imagine that a dotted line is drawn around the object, and note how the line would cut through and expose various forces. Everything within the dotted line is isolated from the surroundings and should appear on the diagram.

2. The coordinate system is drawn next to indicate the positive sign conventions for forces and moments. It will be meaningless to report an answer of, say, "-25 N·m" or "$+250$ lb" without having defined the directions associated with the positive and negative signs.

3. In the final step, all forces and moments are drawn and labeled. These forces might represent weight or contact between the free body and other objects that were removed when the body was isolated. When a force is known, its direction and magnitude should be written on the diagram. Forces are included even if their

magnitudes and directions are not known at this step in the analysis. If the direction of a force is unknown (for instance, upward/downward or leftward/rightward), you should just draw it one way or the other on the FBD, perhaps using your intuition as a guide. After applying the equilibrium equations and consistently using a sign convention, the correct direction will be determined through your calculations. If you find the quantity to be positive, then you know that the correct direction was chosen. On the other hand, if the numerical value turns out to be negative, the result simply means that the force acts opposite to the assumed direction.

EXAMPLE 4.5

During crash testing of an automobile, the lap and shoulder seat belts each become tensioned to 300 lb, as shown in Figure 4.15(a). Treating the buckle B as a particle, (a) draw a free body diagram, (b) determine the tension T in the anchor strap AB, and (c) determine the angle at which T acts.

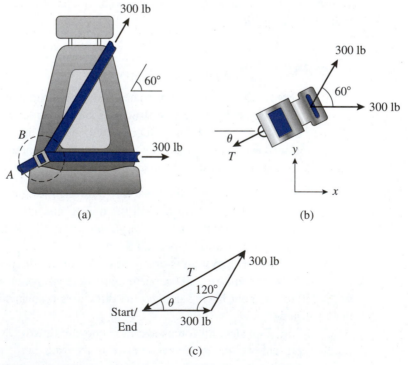

FIGURE 4.15 Equilibrium analysis of the seat belt latch in Example 4.5.

SOLUTION

(a) The free body diagram of the buckle is shown in Figure 4.15(b). The xy coordinate system is also drawn to indicate our sign convention for the positive horizontal and vertical directions. Three forces act on the buckle: the two given 300-lb forces and the unknown force in the anchor strap. For the buckle to be in equilibrium, these three forces must balance. Although both the magnitude T and direction θ of the force in strap AB are unknown, both quantities are shown on the free body diagram for completeness.

(b) We sum the three forces by using the vector polygon approach, as shown in Figure 4.15(c). The polygon's start and end points are the same because the three forces acting together have zero resultant; that is, the distance between the polygon's start and end points is zero. The tension is determined by applying the law of cosines (equations for oblique triangles are reviewed in Appendix B) to the side–angle–side triangle in Figure 4.15(c):

$$T^2 = (300 \text{ lb})^2 + (300 \text{ lb})^2 - 2(300 \text{ lb})(300 \text{ lb}) \cos 120°$$

from which we calculate $T = 519.6$ lb.

(c) The anchor strap's angle is found from the law of sines:

$$\frac{\sin \theta}{300 \text{ lb}} = \frac{\sin 120°}{519.6 \text{ lb}}$$

and $\theta = 30°$. ∎

EXAMPLE 4.6

A pair of wire cutters is shown in Figure 4.16. A machinist applies the 70-N gripping force to the handles. What is the magnitude of the cutting force on the electrical wire at A?

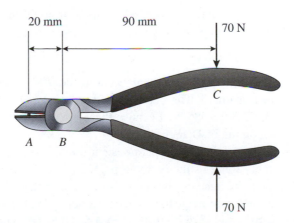

FIGURE 4.16 Equilibrium analysis of the wire cutters in Example 4.6.

SOLUTION

The coordinate system and positive sign convention for the force and moment directions are shown in Figure 4.17. We next draw the free body diagram for one jaw/handle assembly, which is treated as a rigid body because it can rotate and the distances between the forces are significant to the problem. When a blade is pressed against the wire, the wire in turn pushes back with the (unknown) force A on the blade. Force B is carried by the hinge pin, which connects the two jaw/handle pieces. The 70-N gripping force is given and shown on the FBD acting at the end of the handle.

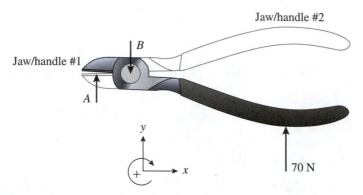

■ FIGURE 4.17 Free body diagram of a jaw/handle in Example 4.6.

With the free body diagram completed in Figure 4.17, the cutting force A is found by applying the equilibrium equations for a rigid body. The force balance requirement in the vertical direction becomes

$$A - B + (70\ \text{N}) = 0$$

There are two unknowns, A and B, and so an additional equation is needed. By summing moments about point B, we have

$$-(70\ \text{N})(90\ \text{mm}) + A(20\ \text{mm}) = 0$$

The negative sign is present because the 70-N force produces a counterclockwise moment about B. The cutting force is found to be $A = 315$ N, and after back substitution, $B = 385$ N. Because these are each positive, the directions assumed and shown on the FBD are correct.

These wire cutters operate according to the principle of a lever. Each jaw/handle assembly rotates slightly about point B. The cutting force at A is proportional to the force on the handles, and it is also related to the ratio of distances AB and BC. The mechanical advantage for a machine is defined as the ratio of the output and input forces, or in this case, $(315\ \text{N})/(70\ \text{N}) = 4.5$. Thus, the wire cutters amplify the machinist's gripping force by some 450%. ■

EXAMPLE 4.7

The forklift shown in Figure 4.18(a) weighs 3500 lb and carries an 800-lb shipping container. There are two front wheels and two rear wheels on the forklift. (a) Draw a free body diagram of the forklift. (b) Determine the contact forces between the wheels and ground. (c) How can you determine the front wheel force without having to find the force at the rear wheels? (d) How heavy a load can be carried before the forklift will start to tip about its front wheels?

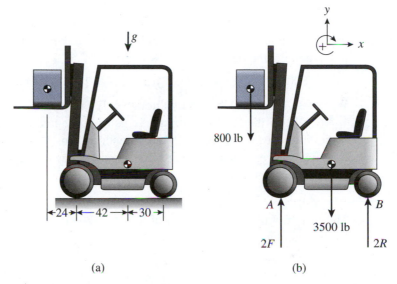

(a)   (b)

■ **FIGURE 4.18**   (a) Geometry and (b) free body diagram of a forklift in Example 4.7. All dimensions are given in inches.

SOLUTION

(a) The free body diagram of the forklift is drawn in Figure 4.18(b), and the positive sign conventions for the forces and moments are also shown. We choose the upward direction as being positive for forces and the clockwise direction positive for moments. We first draw and label the known 3500-lb and 800-lb weights that act through the mass centers of the forklift and container. The (unknown) force between a front wheel and the ground is denoted by F, and the (unknown) force between a rear wheel and the ground is R. On the side view drawing of the FBD, the net effects of those forces on the wheel pairs become $2F$ and $2R$.

(b) There are two unknowns (F and R), and therefore two independent equilibrium equations are needed to solve the problem. We first sum forces in the vertical direction

$$-(800 \text{ lb}) - (3500 \text{ lb}) + 2F + 2R = 0$$

but a second equation is required to solve the problem. Summing forces in the horizontal direction will not provide any useful information, so we use a moment balance. Any location can be chosen as the moment pivot point. We recognize that by choosing the moment point to coincide with the front wheel, force F will be eliminated from the calculation. Thus, taking moments about point A, we have

$$-(800 \text{ lb})(24 \text{ in.}) + (3500 \text{ lb})(42 \text{ in.}) - (2R)(72 \text{ in.}) = 0$$

from which we find $R = 888$ lb. Here, the 800-lb force and the rear wheel forces exert counterclockwise or negative moments about A, and the forklift's 3500-lb weight generates a positive moment about A. Substituting the solution for R into the force balance returns $F = 1262$ lb.

(c) To find the force under the front wheels without having to find R, we can sum moments about the rear wheel at point B. The unknown force R passes through that point, and the moment balance now reads

$$-(800 \text{ lb})(96 \text{ in.}) - (3500 \text{ lb})(30 \text{ in.}) + (2F)(72 \text{ in.}) = 0$$

We obtain $F = 1262$ lb directly through only one equation.

(d) When the forklift is on the verge of tipping about the front wheels, the rear wheel has just lost contact with the ground, and so $R = 0$. We denote as W the new (unknown) weight of the shipping container that causes tipping. The moment balance about the front wheels is now

$$-(W)(24 \text{ in.}) + (3500 \text{ lb})(42 \text{ in.}) = 0$$

The forklift will be on the verge of tipping when the operator attempts to lift a $W = 6125$-lb container.   ∎

4.5 BUOYANCY, DRAG, AND LIFT FORCES IN FLUIDS

Up to this point, we have considered mechanical systems where the forces are associated with gravity or the connections between components. Forces are also generated by stationary and moving fluids, which can be either liquids or gases. In this section, we will examine the three types of fluid forces known as buoyancy, drag, and lift, and along the way, we highlight properties and interesting characteristics of fluids.

The forces generated by liquids and gases are important to many areas of mechanical engineering. Some applications include automobile wind resistance and fuel economy; the flight performance of rockets and aircraft; hydraulic systems in construction machinery; the interactions among the atmosphere, oceans, and the global climate; and even biomechanical studies of air and blood in the human pulmonary system. In Figure 4.19, four forces are shown acting on the aircraft in flight: the plane's weight W, the thrust T produced by the engines, the lift L produced by the wings, and the drag D that opposes motion of the plane through the air.

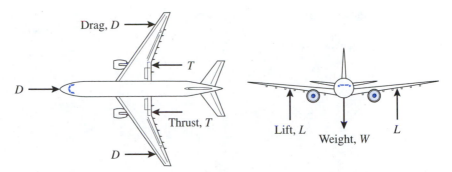

FIGURE 4.19   The weight, thrust, lift, and drag forces acting on an airplane.

Wind tunnels   Wind tunnels, such as those shown in Figure 4.20, are important research and development tools for measuring the forces generated when air flows around an object. Extensively used in the aerospace industry, wind tunnels enable engineers to optimize the performance of aircraft, missiles, and rockets at different speeds and flight conditions. In such a test, a scale model is built and attached to a special fixture for measuring the forces developed by the airstream. As an example, a scale

FIGURE 4.20   An aerial view of several wind tunnels used for aircraft and flight dynamics research.
Source: Reprinted with permission of NASA.

FIGURE 4.21 A 1% scale model of the space shuttle system during ascent. The engineering model comprises the orbiter, two solid rocket engines, the external fuel tank, and exhaust plumes. It is being tested in a transonic wind tunnel.
Source: Reprinted with permission of NASA.

model of the space shuttle system in its ascent configuration is shown during wind tunnel testing in Figure 4.21. That model even includes the exhaust plumes created by the engines on the orbiter and the solid rocket motors, which influence the aerodynamic forces. Wind tunnels are also used to perform experiments for supersonic flight; Figure 4.22 depicts the shock waves that propagate off the scale model of an upper-atmosphere research aircraft. Shock waves occur when the speed of air flowing around an aircraft exceeds the speed of sound, and they are responsible for the noise known as a sonic boom. Wind tunnels are also used to design automobiles that have reduced wind resistance and therefore better fuel economy. Low-speed wind tunnels are applied even in the realm of Olympic sports, to help ski jumpers improve their form and to help engineers design bicycles, cycling helmets, and sporting apparel with improved aerodynamics.

Shock waves

Buoyancy and Pressure

Drag and lift forces, which we will discuss shortly, arise because an object moves through a fluid, or conversely, because a fluid flows past an object. On the other hand, forces between fluids and structures can arise even if they are both stationary, owing to gravity and the weight of the fluid. As you swim to the bottom of a pool or travel in the mountains, the pressure changes in the water or air that surrounds you, and your ears "pop" when adjusting to the rising or falling pressure. Our experience is that the pressure in a liquid or gas increases with depth.

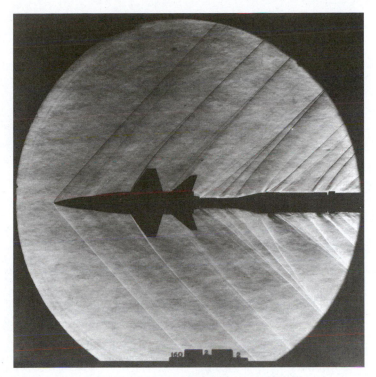

■ FIGURE 4.22   During testing in a supersonic wind tunnel, shock waves propagate off the nose and flight surfaces of the scale model of a research aircraft.
Source: Reprinted with permission of NASA.

Fluid density
- - - - - - - - - - - - →

Referring to the beaker of liquid shown in Figure 4.23, the difference in pressure p between levels "0" and "1" arises because of the liquid's weight. With the two levels separated by depth h, the mass of the liquid column is $m = \rho A h$. Here, ρ (the lowercase Greek character rho) is the liquid's density and Ah is the volume between the two levels. Using the free body diagram of Figure 4.23, the equilibrium force balance for the column reads $p_1 A - p_0 A - (\rho A h)g = 0$, where $g = 9.81 \text{ m/s}^2 = 32.2 \text{ ft/s}^2$ is the gravitational acceleration constant. The pressure in the liquid at depth "1" becomes

$$p_1 = p_0 + \rho g h \tag{4.12}$$

and it increases in direct proportion to depth. Table 4.3 provides numerical values for the density of several common gases and liquids, and Table 4.4 reviews conversion factors between the units for density in the USCS and SI.

Pressure
- - - - - - - - - - - - →

Pascal
- - - - - - - - - - - - →

Pressure has the unit of force-per-unit-area. In the SI, the unit of pressure is the pascal (1 Pa = 1 N/m^2), named after the seventeenth- century scientist Blaise Pascal, who conducted experiments on air and other gases. The units of psi = lb/in^2 and psf = lb/ft^2 are often used for pressure in the USCS, as is the unit of atmosphere or atm. Table 4.5 provides conversion factors

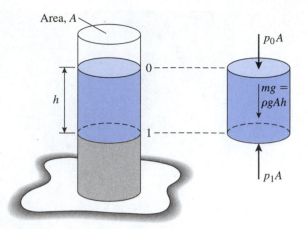

■ **FIGURE 4.23** Equilibrium of a beaker filled with liquid.

TABLE 4.3 Density and Viscosity of several Gases and Liquids at Room Temperature and Pressure

	Density, ρ		Viscosity, μ	
Fluid	**kg/m^3**	**slug/ft^3**	**kg/(m·s)**	**slug/(ft·s)**
Air	1.20	2.33×10^{-3}	1.8×10^{-5}	3.8×10^{-7}
Helium	0.182	3.53×10^{-4}	1.9×10^{-5}	4.1×10^{-7}
Freshwater	1000	1.94	1.0×10^{-3}	2.1×10^{-5}
Seawater	1026	1.99	1.2×10^{-3}	2.5×10^{-5}
Gasoline	680	1.32	2.9×10^{-4}	6.1×10^{-6}
SAE 30 oil	917	1.78	0.26	5.4×10^{-3}

TABLE 4.4 Conversion Factors between USCS and SI Units for Density

kg/m^3	**slug/ft^3**
1	1.940×10^{-3}
515.5	1

between these conventional units. Referring to the table's fourth row, for instance, we see that 1 atm $= 1.013 \times 10^5$ Pa $= 14.70$ psi $= 2116$ psf.

Buoyancy force
$\dashrightarrow$
When ships sail out of port or hot air balloons hover above the ground, they experience buoyancy forces that are created by fluid pressure. The buoyancy force on an object immersed in a fluid develops because the object has

TABLE 4.5   Conversion Factors Between USCS and SI Units for Pressure

Pa (N/m²)	psi (lb/in²)	psf (lb/ft²)	atm
1	1.450×10^{-4}	2.089×10^{-2}	9.869×10^{-6}
6895	1	144	6.805×10^{-2}
47.88	6.944×10^{-3}	1	4.725×10^{-4}
1.013×10^{5}	14.70	2116	1

displaced some volume of fluid, as shown by the submarine in Figure 4.24. The buoyancy force B equals the weight of fluid that is displaced by the object, according to

$$B = \rho_{\text{fluid}} g V_{\text{object}} \tag{4.13}$$

where g is the gravitational acceleration constant, V is the object's volume, and ρ is the fluid's density. Historically, this result is attributed to Archimedes, who is said to have uncovered a fraud in the manufacture of a golden crown that had been commissioned by King Hieros II. The king suspected that an unscrupulous goldsmith had replaced some of the crown's gold with silver. Archimedes recognized that the principle (4.13) could be used to determine whether the crown had been produced from pure gold, or from a less valuable and less dense alloy of gold and silver (see Problem 26 at the end of the chapter).

Drag Force and Viscosity

Whereas a buoyancy force acts in even a stationary fluid, drag force (and the lift force that we discuss in the next section) arises from the motion of an object through fluid (for instance, an automobile and air) or from the flow of fluid past an object (wind loading against a downtown skyscraper). The general behavior of moving fluids defines the field of mechanical engineering known as fluid dynamics. Because the general treatment is rather complex from both the physical and mathematical standpoints, we will consider several restricted cases that are useful to fix ideas and introduce some of the issues. With respect to drag force, we will examine the flow of fluid around a small sphere,

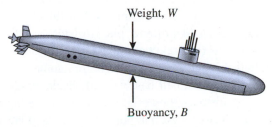

Weight, W

Buoyancy, B

FIGURE 4.24   Buoyancy force on a submerged submarine.

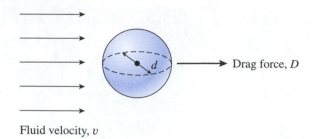

Fluid velocity, v

■ **FIGURE 4.25** Liquid or gas flows past a small sphere, creating the drag force D.

as shown in Figure 4.25. This situation has important applications for devices that deliver medicine through aerosol sprays; to the motion of fine dust, soot, and pollutant particles in the atmosphere; the precipitation of raindrops and hailstones; and the settling of particulate debris from a tank of liquid.

For our purposes, either the sphere moves at speed v through an otherwise stationary liquid or gas or the fluid flows around a stationary sphere. In each case, the drag force D opposes motion and is expressed by

$$D = 3\pi\mu dv \qquad (Re < 1) \tag{4.14}$$

Fluid viscosity
- - - - - - - - - - - - - ->

where d is the sphere's diameter. You can see that the magnitude of the drag force increases in direct proportion to speed and diameter. The parameter μ (the lowercase Greek character mu) in Equation 4.14 is called the fluid's viscosity, and the drag force likewise grows with it. As described in the statement for Problem 9 in Chapter 2, viscosity measures the "stickiness" or resistance of a fluid. Like density, viscosity is a physical property of gases and liquids, and several values are listed in Table 4.3. As you see, in both the SI and USCS, the numerical values for μ are generally small.

Poise
- - - - - - - - - - - - ->

In order for Equation 4.14 to be dimensionally consistent, viscosity must have the units of mass/(length-time). Because the viscosity property arises frequently when fluid systems are analyzed, the special unit of "poise" (P) was created in recognition of the French physician and scientist Jean Poiseuille (1797–1869), who studied blood flow through capillaries in the body. The units of kg/(m·s), slug/(ft·s), and poise are each conventionally used for viscosity, and conversion factors between them are listed in Table 4.6.

Terminal velocity
- - - - - - - - - - - - ->

Equation 4.14 is often applied to determine the speed at which an object shaped like a sphere falls (or rises) through a fluid. Consider the small ball bearing in Figure 4.26 that falls through a container of oil or another liquid. When it is initially dropped into the container, the ball bearing will accelerate downward with gravity. After a short distance, however, it stops accelerating and reaches a constant or terminal velocity. At that point, the drag and buoyancy forces that act upward in the free body diagram of Figure 4.26 exactly balance the ball bearing's weight.

TABLE 4.6   Conversion Factors Between USCS and SI Units for Viscosity

kg/(m·s)	slug/(ft·s)	poise, P
1	0.0209	10
47.88	1	478.8
0.1	2.09×10^{-3}	1

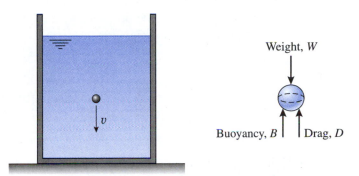

FIGURE 4.26   A ball bearing falls through a tank of oil. The weight, buoyancy, and drag forces balance when the sphere falls at its terminal velocity.

By applying the equilibrium force balance $D + B - mg = 0$, we obtain an expression for the terminal velocity of a small sphere that moves through air, water, or another fluid. By combining Equations 4.13 and 4.14,

$$v_{\text{terminal}} = \frac{gd^2}{18\mu}(\rho_{\text{sphere}} - \rho_{\text{fluid}}) \qquad (Re < 1) \tag{4.15}$$

where g is the gravitational acceleration constant and we have used the expression $\pi d^3/6$ for the volume of a sphere in terms of its diameter. The terminal velocity depends on the difference in density between the solid sphere and the fluid; if the two have the same density, then the sphere has no tendency to either rise or fall. If $v_{\text{terminal}} > 0$, the particle falls downward, but if $v_{\text{terminal}} < 0$, the particle rises in response to the buoyancy force being greater than the sphere's weight. The parenthetical condition $Re < 1$ in both Equations 4.14 and 4.15 imposes a restriction on the values of fluid density, viscosity, speed, and diameter for which these equations are accurate. We discuss that restriction and the dimensionless variable Re, called the Reynold's number, after working through the following examples.

EXAMPLE 4.8

A small water droplet in a mist of air is approximated as a sphere with diameter 1.5 mil. Calculate the terminal velocity as it falls through still air to the ground.

SOLUTION

We use Equation 4.15 to calculate the terminal velocity. Referring to Table 2.2, 1 mil is equivalent to 0.001 inches. However, we must convert the diameter to the units of feet in order to be consistent with the density and viscosity values of Table 4.3. The droplet's diameter is $(0.0015 \text{ in.})(0.0833 \text{ ft/in.}) = 1.25 \times 10^{-4}$ ft. To simplify the calculation somewhat, we decide to make the (very good) approximation that the density of air is much less than that of water. Because the buoyancy force exerted by the air on the droplet is small compared to the droplet's weight, we can neglect the term ρ_{fluid} in Equation 4.15. To a good approximation, the terminal velocity is

$$v \approx \frac{gd^2 \rho_{\text{sphere}}}{18\mu}$$

$$\approx \frac{(32.2 \text{ ft/s}^2)(1.25 \times 10^{-4} \text{ ft})^2(1.94 \text{ slug/ft}^3)}{18(3.8 \times 10^{-7} \text{ slug/(ft·s)})}$$

$$\approx 0.143 \text{ ft/s}$$

or 1.71 in./s, where we have used the density of water and viscosity of air from Table 4.3.   ∎

EXAMPLE 4.9

An experimental engine oil has density 900 kg/m^3, and its viscosity is being measured in a laboratory. A 1-mm-diameter steel ball bearing is released into a much larger, and transparent, tank of the oil (Figure 4.26). After the ball has fallen through the oil for a few seconds, it reaches terminal velocity. A technician records that the ball takes 9 seconds to pass marks on the container that are separated by 10 cm. Knowing that the density of steel is 7830 kg/m^3, what is the oil's viscosity?

SOLUTION

We use Equation 4.15 to determine μ since all other parameters are known. The terminal velocity of the ball bearing is $v = (0.10 \text{ m})/(9 \text{ s}) = 0.0111$ m/s, and direct substitution provides

$$\mu = \frac{gd^2}{18v}(\rho_{\text{sphere}} - \rho_{\text{fluid}})$$

$$= \frac{(9.81 \text{ m/s}^2)(0.001 \text{ m})^2}{18(0.0111 \text{ m/s})}(7830 - 900 \text{ kg/m}^3)$$

$$= 0.34 \text{ kg/(m·s)}$$

In this example, we did not neglect the density of the fluid because it is not small compared to that of the bearing. ∎

Laminar and Turbulent Fluid Flows

Experiments show that Equations 4.14 and 4.15 start to underestimate the drag force as the speed v, diameter d, or the ratio ρ/μ of the fluid's density to viscosity increases. One reason for that breakdown in accuracy is that the analysis behind

Laminar
- - - - - - - - - - - - - - ➤

those equations specifies that the fluid flows smoothly around the sphere, similar to the sketch in Figure 4.27(a). Such a uniform type of fluid motion is called *laminar*, and it is associated with situations where the fluid is moving relatively slowly, the exact definition of "relative" being given shortly. As fluid flows

Turbulent
- - - - - - - - - - - - - - ➤

faster around the sphere, the flow pattern eventually begins to break up and become random-looking, particularly on the sphere's trailing edge. The irregular flow pattern shown in Figure 4.27(b) is said to be *turbulent*.

You may recall an airline pilot instructing you to fasten your seatbelt because of the air turbulence associated with severe weather patterns or mountain ranges; you also have other first-hand experience with laminar and turbulent fluids. Try opening the valve on a kitchen faucet by just a small amount, and watch how water streams out of it in a smooth, glassy-looking, and orderly pattern. The shape of the water stream does not change much from moment to moment, and it might even appear as though the water isn't moving. This is a classic example of laminar water flow. As you gradually open the faucet's valve more and more, you reach a point where the smooth stream of water starts to oscillate, break up, and become turbulent. In general, slowly flowing fluids appear laminar and smooth, but at a high enough speed, the flow pattern becomes turbulent and random-looking.

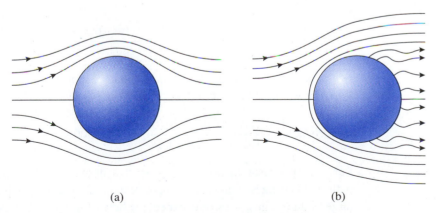

(a)   (b)

■ **FIGURE 4.27**   (a) Laminar and (b) turbulent flow around a sphere.

In the latter half of the nineteenth century, British engineer Osborne Reynolds conducted experiments on the transition between laminar and turbulent flow through a pipe. A dimensionless parameter, now recognized as being the most impor-

Reynolds number
- - - - - - - - - - - - - - ->

tant variable in the engineering analysis of fluids, was found to describe that transition. The Reynolds number, Re, is defined

$$Re = \frac{\rho v l}{\mu} \tag{4.16}$$

Characteristic length
- - - - - - - - - - - - - - ->

in terms of the fluid's density ρ and viscosity μ, speed v, and a length l that is representative of the problem at hand. For crude oil that is being pumped through a pipe, l is the pipe's diameter; for water flowing past the sphere in Figure 4.25, l is the sphere's diameter; for a forced air heating system, l is the width of the air duct; and so forth. The Reynolds number has the interpretation of being the ratio between the inertia and viscous forces in a fluid; the former is proportional to density, and the latter to viscosity. When the fluid moves quickly, the fluid is not very viscous or the fluid is very dense, so the Reynolds number will be large, and vice versa. To develop a feel for the magnitude of Re in different physical circumstances, consider the following examples:

- An automobile travels at 25 m/s. The characteristic length is taken as the width $l = 1.4$ m of the vehicle's body. By using values from Table 4.3 for the density and viscosity of air, the Reynolds number is

$$Re = \frac{(1.20 \text{ kg/m}^3)(25 \text{ m/s})(1.4 \text{ m})}{1.8 \times 10^{-5} \text{ kg/(m·s)}} = 2.3 \times 10^6$$

The units in the calculation precisely cancel, verifying that Re is a dimensionless variable, as discussed in Section 2.3. Reynolds numbers can evidently be large in practice.

- A Winchester .30-30 bullet leaves the muzzle of a rifle at 2400 ft/s. Since the bullet's diameter is 0.3 inch,

$$Re = \frac{(2.33 \times 10^{-3} \text{ slug/ft}^3)(2400 \text{ ft/s})(0.3 \text{ in.})(0.0833 \text{ ft/in.})}{3.8 \times 10^{-7} \text{ slug/(ft·s)}} = 3.7 \times 10^5$$

which is an order of magnitude smaller than Re for the automobile.

- Water flows through a 1-cm-diameter pipe at 0.5 m/s. The Reynolds number is

$$Re = \frac{(1000 \text{ kg/m}^3)(0.5 \text{ m/s})(0.01 \text{ m})}{1.0 \times 10^{-3} \text{ kg/(m·s)}} = 5000$$

Laboratory measurements have shown that fluid flows through a pipe in a laminar pattern when Re is smaller than approximately 2200 and that the flow is turbulent for larger values. Thus, we would expect the flow of water within the pipe to be irregular and turbulent in this case. On the other hand, if SAE 30 oil were pumped through the pipe instead of water,

$$Re = \frac{(917 \text{ kg/m}^3)(0.5 \text{ m/s})(0.01 \text{ m})}{0.26 \text{ kg/(m·s)}} = 18$$

and the flow would certainly be laminar because the oil is so much more viscous than water.

- A fast attack submarine with hull diameter 33 ft cruises at 15 knots (1 knot = 1.152 mph). We convert the speed as $v = 25.4$ ft/s to be consistent with the density and viscosity values for seawater that are tabulated in Table 4.3. The boat's Reynolds number is

$$Re = \frac{(1.99 \text{ slug/ft}^3)(25.4 \text{ ft/s})(33 \text{ ft})}{2.5 \times 10^{-5} \text{ slug/(ft·s)}} = 6.7 \times 10^7$$

Coefficient of Drag

Because the fundamental character of a fluid's flow pattern changes from laminar to turbulent with Re, Equations 4.14 and 4.15 are applicable only when Re is less than about 1. For larger values of Re that can encompass either laminar or turbulent flow around the sphere shown in Figure 4.25, the drag force is determined from the resistance law

$$D = \frac{1}{2}\rho A v^2 C_D \tag{4.17}$$

Frontal area
---------------→

where the area A of the object facing the flowing fluid is called the reference, frontal, or projected area. The parameter C_D is called the coefficient of drag, and it depends on both the shape of the object and the Reynolds number. Numerical values for the drag coefficient are generally determined from laboratory measurements or computer simulations, and they are tabulated and available in the engineering literature for a wide range of situations. The drag coefficient is a single parameter that represents the complex dependency of the drag force on the shape of an object, its inclination angle relative to the flowing fluid, and other effects.

Figure 4.28 presents the drag coefficient for a smooth sphere as a function of Reynolds numbers over the range $0.1 < Re < 100,000$. At the higher values, $1000 < Re < 100,000$, C_D is nearly constant at the value 0.5. By slight algebraic manipulation, we can compare the drag force predicted from Equations 4.14 and 4.17 with the data of Figure 4.17. The drag force found from Equation 4.14 is equivalent to having a drag coefficient of

$$C_D = \frac{24}{Re} \qquad (Re < 1) \tag{4.18}$$

This result is shown as the straight dotted line in the logarithmic representation of Figure 4.28. You can see that the result (4.18) agrees with the more general C_D curve only for Re less than about 1.

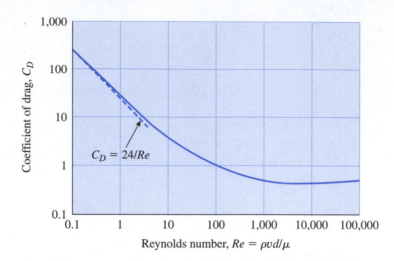

FIGURE 4.28   General dependence of the drag coefficient for a smooth sphere on the Reynolds number (solid line), and the value predicted for low *Re* from Equation 4.18.

EXAMPLE 4.10

Verify that the terminal velocity obtained in Example 4.5 is low enough so that $Re < 1$ and Equation 4.15 was properly applied.

SOLUTION

By using the terminal velocity of 0.011 m/s and the measured viscosity value, we calculate the Reynolds number to be

$$Re = \frac{(900 \text{ kg/m}^3)(0.011 \text{ m/s})(0.001 \text{ m})}{0.34 \text{ kg/(m·s)}} = 0.03$$

Because this is less than 1, we confirm that the equation was properly applied. Had we found that Re was much larger because either the terminal velocity was greater or the viscosity was lower, we would have needed to discard the prediction from Equation 4.15. Equation 4.17 combined with Figure 4.28 would have to be used instead.   ∎

EXAMPLE 4.11

A 1.68 in.-diameter golf ball is driven off a tee at 70 mph (103 ft/s). Determine the drag force acting on the golf ball by (a) approximating it as a smooth sphere, and (b) using the measured coefficient of drag $C_D = 0.27$ that accounts for the dimple pattern on the ball's surface.

SOLUTION

(a) With the diameter of $d = 1.68$ in. $= 0.14$ ft, the Reynolds number is

$$Re = \frac{(2.33 \times 10^{-3} \text{ slug/ft}^3)(103 \text{ ft/s})(0.14 \text{ ft})}{3.8 \times 10^{-7} \text{ slug/(ft·s)}} = 8.8 \times 10^4$$

Referring to Figure 4.28, that value lies in the flat portion of the drag coefficient curve where $C_D \approx 0.5$. The frontal area of the golf ball is

$$A = \frac{\pi}{4}(0.14 \text{ ft})^2 = 0.0154 \text{ ft}^2$$

The drag force from Equation 4.17 then becomes

$$D = \frac{1}{2}(2.33 \times 10^{-3} \text{ slug/ft}^3)(0.0154 \text{ ft}^2)(103 \text{ ft/s})^2(0.5) = 0.095 \text{ lb}$$

or about a tenth of a pound, comparable to the ball's weight.

(b) The dimples change the way air flows around the golf ball and lower the coefficient of drag so that it travels farther in flight. With $C_D = 0.27$, the drag force is lowered to 0.051 lb. The aerodynamic behavior of golf balls is also significantly influenced by any spin that the ball might have. Spin can provide extra lift force (which we discuss next) and enable the ball to travel farther than would otherwise be possible, even in a vacuum. ∎

Lift Force

Similar to drag force, lift is also produced as fluid flows around a structure. While the drag force developed by a fluid acts parallel to the direction of its flow, the lift force

FIGURE 4.29 Mechanical engineers use sophisticated computer analyses to simulate three-dimensional airflow around automobiles and to predict the aerodynamic drag and lift forces.

Source: Reprinted with permission of Fluent Inc.

acts perpendicular to it. In the context of the airplane shown in Figure 4.19, for instance, high-speed flow of air around the wings creates vertical lift that balances the plane's weight during flight. Lift force is important not only for aircraft wings and other flight control surfaces but also for the design of propeller, compressor, and turbine blades; ship hydrofoils; and body contours of commercial and racing automobiles (Figures 4.29–4.32).

Aerodynamics
- - - - - - - - - - - - - - - ➤

The area of mechanical engineering that encompasses the interaction between structures and the air flowing around them is called *aerodynamics*.

It is an exacting field because of the complex mathematical description for flow over aircraft surfaces and the intricate nature of the many physical processes involved. The engineering analysis of drag, lift, and other aspects of fluid motion invariably involves approximations of geometry and fluid properties. For that

■ **FIGURE 4.30**   An F-22 Raptor fighter climbs at a steep 55° angle of attack.
Source: Reprinted with permission of Lockheed-Martin.

■ **FIGURE 4.31**   The space shuttle orbiter Atlantis touches down at the Kennedy Space Center after a mission lasting over ten days. The aerodynamic design of the orbiter is a compromise for operation over a wide range of conditions: ascent through the relatively dense lower atmosphere, hypersonic re-entry in the upper atmosphere, and unpowered gliding flight to touchdown. In that latter phase, the orbiter has been described, only partially in jest, as a "flying brick."
Source: Reprinted with permission of NASA.

reason, engineering investigations are often based on a combination of analyses, laboratory experiments, and empiricism. The approximations of neglecting viscosity or compressibility, for instance, might simplify a certain situation enough for an engineer to prepare a preliminary design or interpret measurements. On the other hand, engineers are mindful of the fact that while those approximations are meaningful in some applications, they would be inappropriate in others. As in Equations 4.14 and 4.15, mechanical engineers should be aware of the restrictions involved when equations are applied. For instance, an approximation that we are implicitly making in this chapter involves viewing air as continuous fluid, and not as a collection of discrete molecules that collide with one another. That treatment is very good for most everyday applications involving the flow of air around automobiles and of aircraft at low speeds and altitudes. On the other hand, for aircraft or space vehicles in the upper atmosphere, that approximation might not be appropriate, and engineers and scientists could instead examine fluid forces from the standpoint of the kinetic theory of gases.

Airfoil

----------------►

In a manner similar to our treatment of drag force, the lift force acting on a wing or airfoil is quantified by the coefficient of lift, C_L, a single parameter that groups together the effects of various shape and flow phenomena.

■ **FIGURE 4.32**   Mechanical engineers use computer-aided engineering tools to simulate (a) the air pressure around a helicopter in forward flight and (b) the irregular three-dimensional flow of air around the rotor and airframe.
Source: Reprinted with permission of Fluent Inc.

Lift coefficient

Angle of attack

The amount of lift that is generated by an airfoil depends on its shape, speed, and tilt or angle of attack α relative to the air stream. The flow past the airfoil of Figure 4.33 produces the vertical lift force L, which results from the pressure difference between the air moving past the airfoil's lower surface and the air moving along its upper surface. We discuss that process further in Chapter 6. Analogous to Equation 4.17, lift force is calculated from the expression

$$L = \frac{1}{2}\rho A v^2 C_L \tag{4.19}$$

Chord area

Numerical values for the lift coefficient of a thin airfoil and one having a symmetric cross section are shown in Figure 4.34 as a function of α. In Figure 4.33, $A = ab$ is called the *chord area*, and it is measured across the airfoil's central plane.

Mach number

Meaningful application of Equation 4.19 is restricted to situations of low speed and low angle of attack. As described in Section 2.3, the Mach number Ma is a dimensionless parameter defined as the ratio of the aircraft's speed to the speed of sound in air, which is approximately 700 mph. At speeds greater than

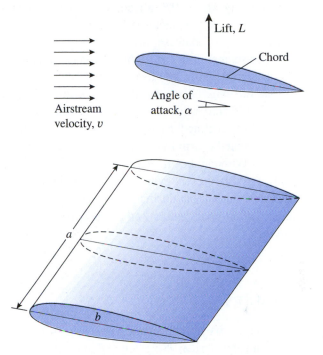

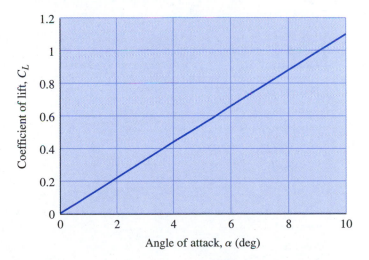

■ **FIGURE 4.33**   Vertical lift force is created as fluid flows past an airfoil inclined at the angle of attack α.

■ **FIGURE 4.34**   Lift coefficient as a function of the angle of attack for a thin symmetric airfoil.

approximately $Ma = 0.3$, air compresses by a significant amount as it flows around airfoils, and the lift characteristics change relative to those described by Equation 4.19 and Figure 4.34. Likewise, for α outside the range of Figure 4.34, the lift coefficient can decrease, resulting in a flight phenomenon known as stall. Both of those effects are not considered by our first level of analysis.

One attribute of an airfoil that is symmetric about its centerline is that the lift coefficient is zero when $\alpha = 0$. That is, when air flows parallel to the airfoil, no lift

Camber
------------------------→

is generated. Real aircraft wings generally have some amount of camber so that the airfoil's centerline is curved slightly with a concave downward shape.

With camber, the curve of Figure 4.34 shifts upward so that a lift coefficient of, say, 0.25, is present even with zero angle of attack. Aircraft wings often are only slightly cambered, but during low-speed flight at takeoff and landing where stall is a concern, additional camber can be created by extending flaps on the trailing edge of the wings.

SUMMARY

In this chapter, we introduced the engineering science concepts of forces, resultants, moments, and equilibrium. We examined those quantities in the context of forces acting on machines and structures and their interactions with a fluid. The primary variables, symbols, and conventional units that are used in this chapter are summarized in Table 4.7.

TABLE 4.7   Quantities, Symbols, and Units That Arise When Treating Forces in Structures and Fluids

Quantity	Conventional Symbols	Conventional Units	
		USCS	SI
Force component	F_x, F_y	lb	N
Resultant	R	lb	N
Buoyancy force	B	lb	N
Drag force	D	lb	N
Lift force	L	lb	N
Moment	M_o	in·lb, ft·lb	N·m
Pressure	p	psi, psf	Pa
Density	ρ	slug/ft^3	kg/m^3
Viscosity	μ	slug/(ft·s)	kg/(m·s)
Sphere diameter	d	mil, in.	μm, mm

TABLE 4.7 *(continued)*

Quantity	Conventional Symbols	Conventional Units	
		USCS	SI
Characteristic length	l	ft	m
Frontal or chord area	A	ft^2	m^2
Velocity	v	ft/s	m/s
Reynolds number	Re	—	—
Coefficient of drag	C_D	—	—
Coefficient of lift	C_L	—	—
Mach number	Ma	—	—

You should note that the equations in this chapter in and of themselves are not particularly complicated from the algebraic standpoint. The skill that mechanical engineers develop, however, is the ability to apply those equations to physical problems in a clear and consistent manner. Selecting the object to be included in a free body diagram, choosing the directions for coordinate axes, and picking the best point for taking moments are examples of some choices that you need to make when solving engineering problems. In addition, the validity of some equations is restricted, and they can be reasonably applied only in certain circumstances. In Chapter 5, we extend our investigation of force systems to include the properties of engineering materials so that we can begin to estimate whether a certain structure or machine will be strong enough to support the forces acting on it.

SELF-STUDY AND REVIEW

1. What are the units for force and moment in the USCS and SI?
2. How many newtons are equivalent to 1 pound?
3. How do you calculate the resultant of a force system by using the vector algebra and vector polygon methods? Give examples of instances where it would be more expedient to use one method over the other.
4. How do you calculate a moment by using the perpendicular lever arm and force components methods? Give examples of instances where it would be more expedient to use one method over the other.
5. Why is a sign convention used when calculating moments using the component method?

6. What are the equilibrium requirements for particles and rigid bodies?
7. What are the steps for drawing a free body diagram?
8. What are the units for fluid density and viscosity in the USCS and SI?
9. Make sketches to illustrate the difference between laminar and turbulent fluid flow.
10. How is the Reynolds number defined?
11. Give examples of buoyancy, drag, and lift forces, and explain how they are calculated.
12. What are the coefficients of drag and lift?
13. What is the definition of the Mach number in aerodynamics?

PROBLEMS

1. Find an example of a real mechanical structure or machine that has several forces acting on it. (a) Make a clear, labeled drawing of the situation. (b) Estimate the dimensions of the structure or machine and the magnitudes and directions of the forces that act on it. Show these on your drawing. Briefly explain why you estimate the dimensions and forces to have the numerical values that you assigned. (c) By using a method of your choice, calculate the resultant of the system of forces.

2. Horizontal and vertical forces act on the pillow block bearing as it supports a rotat-

ing shaft. Determine the magnitude of the resultant force and its angle relative to horizontal. Is the resultant force on the bearing a thrust or radial force?

3. The figure shows a top view of a cylindrical coordinate robot on a factory assembly line. The 50-N force acts on a work piece being held at the end of the robot's arm. Express the 50-N force as a vector in terms of unit vectors **i** and **j** that are aligned with the x and y axes.

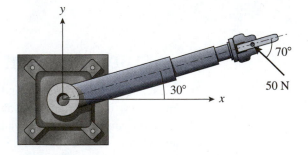

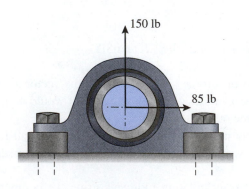

4. During the power stroke of an internal combustion engine, the 400-lb pressure force pushes the piston down its cylinder. Determine the components of that force in the directions along, and perpendicular to, the connecting rod AB.

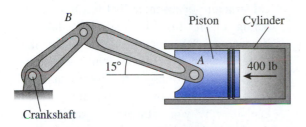

5. A vector polygon for summing 2-kN and 7-kN forces is shown. Determine (a) the magnitude R of the resultant by using the law of cosines and (b) its angle of action θ by using the law of sines.

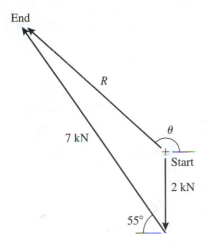

6. A hydraulic lift truck carries a shipping container on the inclined loading ramp in a warehouse. The 12-kN and 2-kN forces act

on a rear tire as shown. (a) Express the resultant of those two forces as a vector using the unit vectors **i** and **j**. (b) Determine the magnitude of the resultant and its angle relative to the incline.

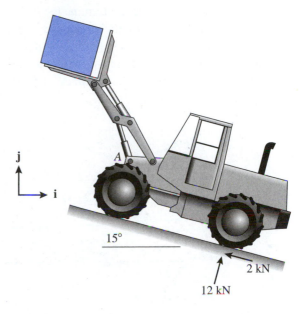

7. Three tension rods are bolted to a gusset plate. Determine the magnitude and direction of

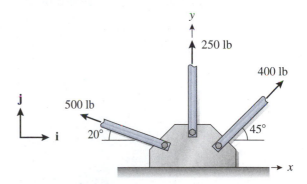

their resultant. Use the (a) vector algebra and (b) vector polygon methods. Compare the answers from the two methods to verify the accuracy of your work.

8. The bucket of an excavator at a construction site is subjected to 1200-lb and 700-lb digging forces at its tip. Determine the magnitude and direction of their resultant. Use the (a) vector algebra and (b) vector polygon methods. Compare the answers from the two methods to verify the accuracy of your work.

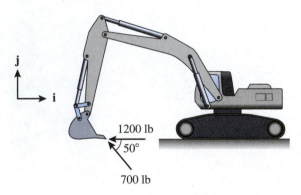

9. Forces of 225 N and 60 N act on the tooth of a spur gear. The forces are perpendicular to one another, but they are inclined by 20°

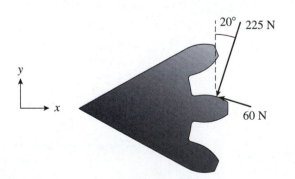

relative to the xy axes. Determine the magnitude and direction of their resultant. Use the (a) vector algebra and (b) vector polygon methods. Compare the answers from the two methods to verify the accuracy of your work.

10. Find a real physical example of a mechanical structure or machine that has a moment acting on it. (a) Make a clear, labeled drawing of the situation. (b) Estimate the dimensions of the structure and the magnitudes and directions of the forces that act on it. Show these on your drawing. Briefly explain why you estimate the dimensions and forces to have the numerical values that you assigned. (c) Select a moment pivot point, explain why you chose that point (perhaps it is important from the standpoint of the structure or machine not breaking), and estimate the moment produced about that point.

11. Resulting from a light wind, the air pressure imbalance of 100 Pa acts across the

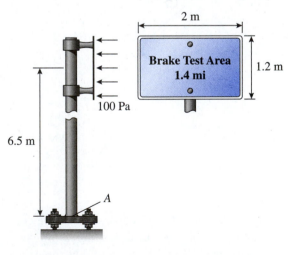

1.2-m by 2-m surface of the highway sign. (a) Calculate the magnitude of the force acting on the sign. (b) Calculate the moment produced about point A on the base of the pole.

12. The spur gear has a pitch radius of 2.5 in. During operation of a geartrain, a 200-lb meshing force acts at 25° relative to horizontal. Determine the moment of that force about the center of the shaft. Use the (a) perpendicular lever arm and (b) moment components methods. Compare the answers of the two methods to verify the accuracy of your work.

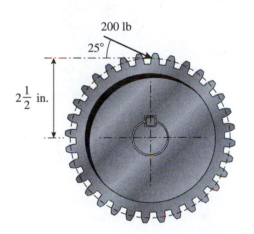

200 lb

25°

$2\frac{1}{2}$ in.

13. Determine the moment of the 35-lb force about the center A of a hex nut. Use the moment components method.

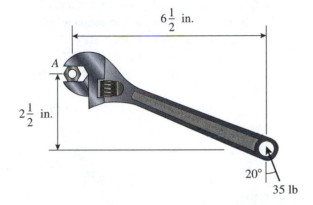

$6\frac{1}{2}$ in.

A

$2\frac{1}{2}$ in.

20°

35 lb

14. Two construction workers pull on the control lever of a frozen steam valve. The lever connects to the valve's stem through the key that fits into partial square grooves on the shaft and handle. Determine the net moment about the center of the shaft.

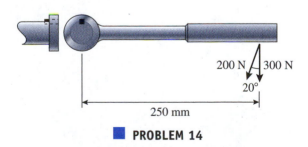

200 N   300 N

20°

250 mm

■ PROBLEM 14

15. Gripper C of the articulated-joint industrial robot is accidentally subjected to a 60-lb side load, directed perpendicular to BC. The robot's link lengths are $AB = 22$ in. and $BC = 18$ in. By using the moment components method, determine the moment of this force about the center of joint A.

moment of this force about the lower support point C of the boom.

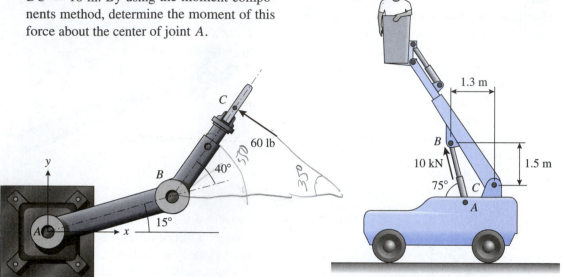

16. A mobile boom lift is used in construction and maintenance applications. The hydraulic cylinder AB exerts a 10-kN force, directed along the cylinder, on joint B. By using the moment components method, calculate the

In problems 17–25, be sure to draw clear free body diagrams and indicate the positive directions for forces or moments.

17. A trough of concrete weighs 800 lb. (a) Draw a free body diagram of cable ring A.

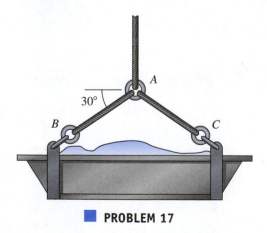

■ **PROBLEM 17**

(b) Treating the cable ring as a particle, determine the tension in cables AB and AC.

18. Cable AB of a boom truck is used to hoist the 2500-lb section of precast concrete. A second cable has tension P, and workers use it to pull and adjust the position of the section as it is being raised. (a) Draw a free

body diagram of hook A, treating it as a particle. (b) Determine P and the tension in cable AB, which is inclined 5° from vertical.

19. Solve the problem of Example 4.5 by using the force component method. Replace the polar representation of the anchor strap's tension by the horizontal and vertical components T_x and T_y, and solve for them. Use your solution for T_x and T_y to determine the magnitude T and direction θ of the anchor strap's tension.

20. A front loader with mass 4.5 Mg is shown in side view as it lifts a 0.75-Mg load of gravel. (a) Draw a free body diagram of the front loader. (b) Determine the contact forces between the wheels and the ground. (c) How heavy a load can be carried before the loader will start to tip about its front wheels?

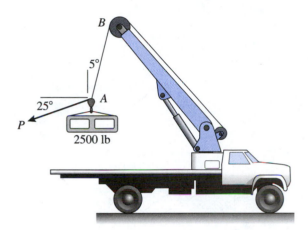

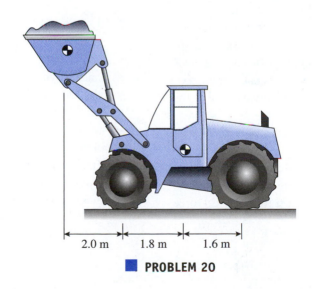

PROBLEM 20

21. Channel lock pliers hold a round metal bar as a machinist grips the handles with $P = 50$ N. Using the free body diagram shown for the combined lower jaw and upper handle, calculate the force A that is applied to the bar.

on the upper jaw. (a) Complete the free body diagram of the upper jaw. (b) Determine the cutting force F that is applied to the pipe.

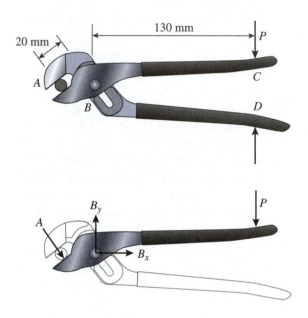

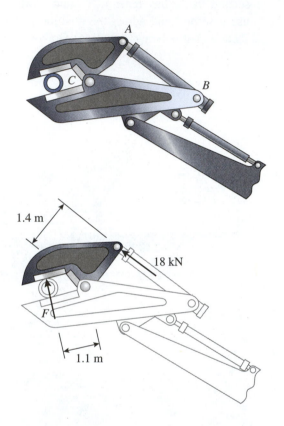

22. For the exercise of Problem 21, (a) measure the angle of force A directly from the diagram and find the magnitude of the force at hinge B. (b) A design condition is that the force at B should be less than 5 kN. What is the maximum force that a machinist can apply to the handle?

23. A pair of large hydraulically operated shears is attached to the end of a boom on an excavator. The shear is used for cutting steel pipe and I-beams during demolition work. Hydraulic cylinder AB exerts an 18-kN force

24. A handrail (see top of next page), which weighs 120 N and is 1.8-m long, was mounted to a wall adjacent to a small set of steps. The support at A has broken, and the rail has fallen about the loose bolt at B so that one end now rests on the smooth lower step. (a) Draw a free body diagram of the handrail. (b) Determine the magnitude of the force at B.

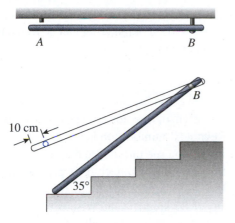

25. A multipurpose utility tool grips a cotter pin at A as 15-lb forces are applied to the handles. (a) Complete the free body diagram of the combined upper jaw and lower handle assembly. (b) Calculate the force acting at A. (c) How much greater would the force be if the pin was being cut at B?

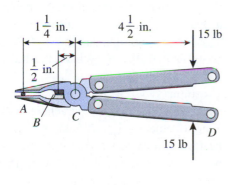

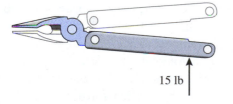

26. An ancient king's "golden" crown had a mass of 3 kg, but it was actually made from an equal mix of gold (1.93×10^4 kg/m^3) and silver (1.06×10^4 kg/m^3) by a dishonest metal smith. (a) Suppose that Archimedes suspended the crown from a string and lowered it into water until it was fully submerged. If the string was then connected to a balance scale, what tension would Archimedes have measured in the string? (b) If the test was repeated, this time with the crown replaced by a 3-kg block of pure gold, what tension would be measured?

27. Scuba divers often carry ballast weights in order for them to have neutral buoyancy. At that condition, the buoyancy force on the diver exactly balances weight, and there is no tendency either to float toward the surface or to sink. In freshwater, a certain diver carries 10 lb of lead alloy (1.17×10^4 kg/m^3) ballast. During an excursion in seawater, the diver must carry 50% more ballast to remain neutrally buoyant. How much does the diver weigh?

28. A steel storage tank is filled with gasoline. The tank has partially corroded on its inside, and small particles of rust have contaminated the fuel. The rust particles are spherical and have a diameter of 25 μm, and density of 5.3 g/cm^3. (a) What is the terminal velocity of the particles as they fall through the gasoline? (b) How long would it take the particles to fall 5 m and settle out of the tank?

29. In a production woodworking shop, 50-μm spherical dust particles were blown into the air while a piece of oak furniture was sanded.

(a) What is the terminal velocity of the particles as they fall through the air? (b) Neglecting any air currents that are present, how long would it take the cloud of sawdust to settle out of the air and fall 2 m to the ground? The density of dry oak is approximately 750 kg/m^3.

30. (a) A 1.5-mm-diameter steel ball bearing (7830 kg/m^3) is dropped into a tank of SAE 30 oil. What is its terminal velocity? (b) If the ball bearing is instead dropped into a different oil of the same density but develops a terminal speed of 1 cm/s, what is the oil's viscosity?

31. A 175-lb skydiver reaches a terminal velocity of 150 mph during free fall. If the frontal area of the diver is 8 ft^2, what are (a) the magnitude of the drag force acting on the skydiver and (b) the drag coefficient?

32. A low-altitude meteorological research balloon, temperature sensor, and radio transmitter together weigh 2.5 lb. When inflated with helium, the balloon is spherical with a diameter of 4 ft. The volume of the transmitter can be neglected when compared to the balloon's size. The balloon is released from ground level and quickly reaches its terminal ascent velocity. Neglecting variations in the atmosphere's density, how long does it take the balloon to reach an altitude of 1000 feet?

33. Examine the transition between laminar and turbulent water flow by making sketches of the stream of water that flows out of a faucet or hose. You can control the velocity of the stream by adjusting the water valve. Make sketches for four different fluid speeds, two below and two above the laminar-turbulent transition point. Estimate the velocity of the water by determining the time Δt required to fill a container of known volume V, such as a plastic soda bottle. By measuring the diameter d of the stream with a ruler, the average speed of the water can be found from the equation

$$v = \frac{4V}{\pi d^2 \Delta t}$$

For each of the four speeds, calculate the Reynolds number. Indicate the Re value where you see turbulence begin.

34. In a pipeline connecting a production oilfield to a tanker terminal, oil having density 1.85 slug/ft^3 and viscosity 6×10^{-3} slug/(ft·s) flows through a 48-in.-diameter pipeline at 6 mph. What is the Reynolds number?

35. Place Equation 4.14 into the form of Equation 4.17 and show that the coefficient of drag for small Reynolds number is given by Equation 4.18.

36. (a) A luxury sports car has a frontal area of 22.4 ft^2 and a 0.29 coefficient of drag at 60 mph. What is the drag force on the vehicle at this speed? (b) A sport utility vehicle has $C_D = 0.45$ at 60 mph and the slightly larger frontal area of 29.1 ft^2. What is the drag force in this case?

37. A certain type of parachute has a drag coefficient of 1.5. If the parachute and skydiver together weigh 225 lb, what should the frontal area of the parachute be so that the skydiver's terminal velocity is 15 mph when approaching the ground? Is it reasonable to neglect the buoyant force that is present?

38. The spoiler on the back of an amateur racing automobile is 4 feet wide and 1.5 feet long, and it is tilted downward by 7° to increase the traction force between the rear wheels and the race course. Calculate the downward force that is generated at a speed of 100 mph.

39. Submarines dive by opening vents that allow air to escape from ballast tanks and water to flow in and fill them. In addition, diving planes located at the bow are angled downward to help push the boat below the surface. Calculate the diving force produced by a 20 ft^2 hydroplane that is inclined by 3° as the boat cruises at 15 knots (1 knot = 1.152 mph).

5 Materials and Stresses

5.1 OVERVIEW

As part of their responsibilities, mechanical engineers must design hardware so that it doesn't break. To that end, by applying the properties of force systems as described in Chapter 4, we can calculate the magnitudes and directions of forces that act on structures and machines. Knowing those forces alone, however, is not sufficient for an engineer to predict whether a certain mechanical component will be strong enough and not fail in the task at hand. By "fail" or "failure" we mean that the hardware will not break, stretch, or bend so much as to become nonusable. A 5-kN force, for instance, might be large enough to break a small bolt or bend a shaft to the point that it would be useless. A larger-diameter shaft or one made from a higher-grade material, on the other hand, might very well be able to support that 5-kN force without any damage.

With those thoughts in mind, you can see that the conditions for a machine component to fail depend not only on the forces that are applied to it but also on its dimensions and the properties of the material from which it is made. As an example, the broken crankshaft shown in Figure 5.1 was salvaged from a single-cylinder internal combustion engine. This particular failure was accelerated by the sharp corners that had been cut in the shaft's rectangular keyway. The spiral shape of the fracture surface indicates that the shaft had been overloaded by a high torque prior to breaking. In short, engineers are able to combine their knowledge of forces, materials, shapes, and dimensions in order to learn from past failures and to better design new hardware.

In this chapter, we will discuss the properties of engineering materials and examine the various types of stresses that can develop within them. Tension, compression, and shear stress are quantities that engineers calculate when they relate the dimensions of a mechanical component to the forces acting on it. Those stresses are then compared with the material's physical properties in order to determine whether or not failure is expected to occur. Engineers generally conduct this type of force, stress, and failure

■ **FIGURE 5.1**   A broken crankshaft from a single-cylinder internal combustion engine.

analysis early in a product's design cycle as choices are made for dimensions, materials, and operating conditions. After completing this chapter, you should be able to:

- Identify circumstances where a machine component is loaded in tension, compression, or shear, and calculate the stress that is present.
- Draw stress–strain curves for aluminum and steel, and determine the elastic modulus and yield strength from such diagrams.
- Understand the differences between elastic and plastic behavior.
- Apply a factor of safety to components that are subjected to tension or shear stress.

5.2   TENSION AND COMPRESSION

The type of stress that is most readily visualized, and therefore most useful for you to develop intuition about, is called *tension* or *compression*. Figure 5.2 shows a round rod that is built in and held fixed at its left end and placed in tension by the force that pulls on its right end. In Figure 5.2(a), the rod has unstretched length L, diameter d, and cross-sectional area $A = \pi d^2/4$. Engineers usually calculate the cross-sectional area of round rods, bolts, and shafts in terms of their diameter, rather than radius (πr^2), because the diameter can be directly measured with a caliper gauge. As the force is gradually applied, the rod stretches along its length by the amount ΔL shown in Figure 5.2(b). In addition, the diameter of the rod shrinks by a small amount due to the effect known as Poisson contraction, a topic that is described further in the next section. However, the change in diameter, Δd, is much smaller and less noticeable than the lengthwise

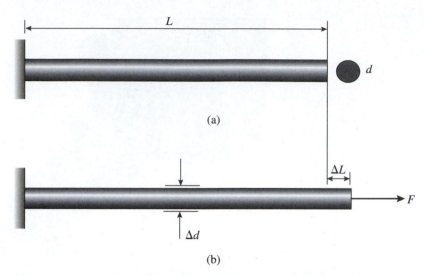

(a)

(b)

■ **FIGURE 5.2**   The straight rod is stretched and placed in tension by force *F*.

stretch ΔL. To gauge the relative amounts of ΔL and Δd, try stretching a rubber band to notice how its length, width, and thickness each change.

In Figure 5.2, the rod will spring back to its original diameter and length when the force is removed. Because the rod does not take on a permanent set after *F* has been applied, the stretching is said to occur elastically. Alternatively,

Elastic behavior
- - - - - - - - - - - - - - - - →

the force could have been large enough that the rod would have been plastically deformed, meaning that when the force was applied, and then removed, the rod would end up being longer than it was initially. You can experiment

Plastic behavior
- - - - - - - - - - - - - - - - →

with a desktop paper clip to see firsthand the difference between the elastic and plastic behavior of materials. Bend one end of the paper clip just a small amount, perhaps a millimeter or two, and notice how it springs back to the original shape when you release it. On the other hand, you can also unwrap the paper clip into a nearly straight piece of wire. In that case, the paper clip does not spring back, and the force was large enough to permanently change shape through plastic deformation.

Although the force is applied at only one end of the rod, its influence is felt at each cross section along the rod's length. As shown in Figure 5.3, imagine slicing through the bar at some point inward of the right-hand end. The segment of rod that is isolated in the free body diagram of Figure 5.3(b) shows *F* acting at its right-hand end

Internal force
- - - - - - - - - - - - - - - - →

and an internal force at the segment's left-hand end that balances *F*. Such must be the case, or otherwise the segment in the free body diagram would not be in equilibrium. The location of our hypothetical slice through the rod is arbitrary, so a force of magnitude *F* must be present at each internal cross section.

Because the rod is formed of a solid continuous material, we do not realistically expect that the internal force remains concentrated at a point as sketched in Figure 5.3(b). Instead, the effect of the force spreads out over the cross section, a process that is

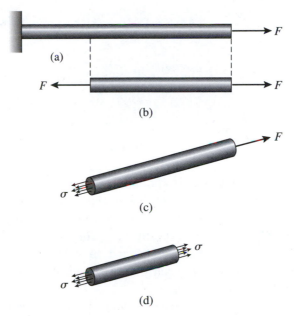

■ FIGURE 5.3 Description of tensile stress in a rod. (a) The stretched rod with end load, (b) a section sliced from the rod exposing the internal force *F*, (c) the stress distributed over the cross section, and (d) tensile stresses acting on the ends of any element sliced from the rod.

the basic idea behind stresses within mechanical components. Stress is essentially an internal force that has been spread over a cross section, as shown in Figure 5.3(c). The rod's tensile stress is defined as the applied force per unit of original cross-sectional area according to the equation

$$\sigma = \frac{F}{A} \tag{5.1}$$

Tension

Compression

The stress σ (the lowercase Greek character sigma) is perpendicular to the slices depicted in Figure 5.3(d). When the stress tends to lengthen the rod, it is called *tension*, and $\sigma > 0$. On the other hand, when the rod would be shortened, the stress is called *compression*. In that case, the direction of σ in Figure 5.3 would reverse to point inward, and $\sigma < 0$.

In Chapter 4, we discussed the pressure acting within a liquid or gas; pressure is also interpreted as a force that is distributed over an area. Stress and pressure therefore have the same units, and in the SI, the unit of stress is the Pascal (1 Pa = 1 N/m^2). The unit psi = lb/in^2 is used for stress in the USCS. Because large numerical quantities frequently arise in the analysis of stress and material properties, the prefixes kilo- (k), mega- (M), and giga- (G) are used to represent factors of 10^3, 10^6, and 10^9, respectively. For instance, 1 kPa = 10^3 Pa, 1 MPa = 10^6 Pa, and 1 GPa = 10^9 Pa. Despite mixing formats with the SI, it is also conventional to use

kPa, MPa, GPa

TABLE 5.1 Conversion Factors Between USCS and SI Units for Stress

psi (lb/in^2)	Pa (N/m^2)
1	6895
1.450×10^{-4}	1

ksi, Mpsi
- - - - - - - - - - - - - - - ➤

the prefixes kilo- and mega- when representing large numbers in the USCS. Mechanical engineers abbreviate the stresses 10^3 psi as 1 ksi (without the "p"), and 10^6 psi as 1 Mpsi (with the "p"). In the USCS, the unit of 1 billion psi (Gpsi) is unrealistically large for calculations involving materials and stresses in mechanical engineering, and so it is not conventionally used. Numerical values for stresses are converted between the USCS and SI using the factors listed in Table 5.1.

Elongation
- - - - - - - - - - - - - ➤

The stretch or elongation ΔL is one way to describe how the rod lengthens when F is applied, but it is not the only way, nor is it necessarily the best way. If a second rod has the same cross-sectional area as in Figure 5.2 but is only half as long, then according to Equation 5.1, the stress within it would be the same as for the first rod. However, we expect intuitively that the shorter rod would stretch by a smaller amount. To convince yourself of this, hang a weight from two different lengths of rubber band, and notice how the longer band stretches more. Just

Strain
- - - - - - - - - - - - - - - ➤

as stress is a measure of force per unit area in the rod, the quantity called *strain* is defined as the amount of stretching that occurs per unit of length. Strain ε (the lowercase Greek character epsilon) is scaled to the rod's original length and calculated from the expression

$$\varepsilon = \frac{\Delta L}{L} \tag{5.2}$$

Because the length units cancel in the numerator and denominator, ε is a dimensionless quantity. Strain is generally very small, and you can express it as a decimal (for instance, $\varepsilon = 0.005$) or as a percent ($\varepsilon = 0.5\%$).

EXAMPLE 5.1

The U-bolt shown in Figure 5.4 is used to attach the body (formed with I-beam construction) of a commercial moving van to its chassis (formed from hollow box channel). The U-bolt is made from a 10-mm-diameter rod, and the nuts on it are tightened until the tension is 4 kN. Calculate the tensile stress in a straight section of the bolt.

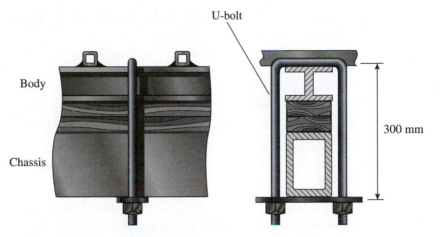

FIGURE 5.4 Stress in the straight shaft section of a U-bolt in Example 5.1.

SOLUTION

The cross-sectional area of the bolt is

$$A = \frac{\pi d^2}{4}$$

$$= 78.5 \text{ mm}^2$$

$$= 7.85 \times 10^{-5} \text{ m}^2$$

The stress then becomes

$$\sigma = \frac{F}{A}$$

$$= \frac{4000 \text{ N}}{7.85 \times 10^{-5} \text{ m}^2}$$

$$= 5.1 \times 10^7 \text{ Pa}$$

$$= 51 \text{ MPa}$$

where we have used the SI prefix for 1 million in order to write the result in a more compact form. ∎

5.3 RESPONSE OF ENGINEERING MATERIALS

The definitions for stress and strain, in contrast to force F and elongation ΔL, are useful because they are scaled with respect to the rod's size. Imagine conducting a sequence of experiments with a collection of rods made of identical material but having various diameters and lengths. As each rod is pulled in tension, its elongation would be recorded.

In general, for a given value of F, each rod would stretch by a different amount because of the differences in diameter and length.

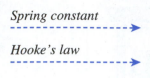

For each individual rod, however, you would find that the applied force F and elongation ΔL are proportional according to $F = k\Delta L$. The parameter k is called the *stiffness* or *spring constant*. This observation is the basis of the concept called *Hooke's law*, which states that force and deflection are proportional. In fact, Robert Hooke wrote in 1678

> … the power of any spring is in the same proportion with the tension thereof; that is, if one power stretch or bend it one space, two will bend it two, and three will bend it three, and so forward.

Note that Hooke used the term *power* for what we today call *force*. Any component that stretches, bends, or deforms elastically can be viewed as a spring having stiffness k, even though the component itself is not necessarily a "spring" in the sense of being a wire helix.

Continuing with our thought experiment, we next construct a graph of F versus ΔL for each of the rods. As indicated in Figure 5.5, the lines on these graphs would have different slopes (or spring constants) depending on d and L for the sample at hand. For a given force, longer rods and ones with smaller cross sections would stretch more than other rods. However, because all of the rods are made from the same material, they would behave in a similar manner when stretching is described instead by stress σ and strain ε, as shown in Figure 5.6. Our conclusion from such experiments is that while the spring constant k depends on the rod's dimensions, the σ–ε relationship is a property of the material alone and independent of the test specimen's size.

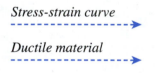

Figure 5.7 shows an idealized stress–strain curve for a material such as ductile low-carbon steel. Ductility refers to the ability of a material to withstand a significant amount of stretching before it breaks. Structural steels and some aluminum alloys are examples of ductile metals, and they are often used in machine components because when they become overloaded, they

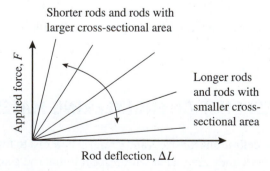

FIGURE 5.5 Force–deflection behavior of rods having various cross-sectional areas and lengths.

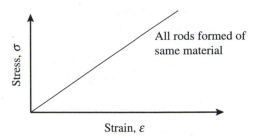

FIGURE 5.6   Each rod has similar stress–strain behavior.

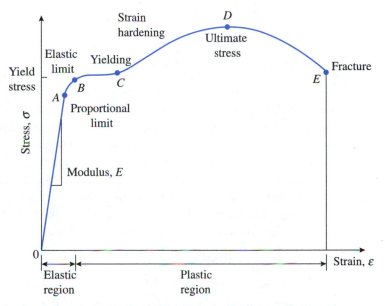

FIGURE 5.7   Idealized stress–strain curve for a ductile low-carbon steel.

Brittle material
- - - - - - - - - - - - →

Elastic and plastic regions
- - - - - - - - - - - - →

Proportional limit
- - - - - - - - - - - - →

tend to noticeably stretch or bend before they break. On the other hand, so-called brittle materials such as cast iron, concrete, ceramics, and glass break suddenly and without much prior warning when they are overloaded.

The stress–strain diagram for a ductile material is broken down into two regions: the low-strain elastic region (where no permanent set remains after a load has been applied and removed) and the higher-strain plastic region (where the load is large enough that, upon removal, the material is permanently elongated). For strains below the proportional limit A in Figure 5.7, stress and strain are linearly related according to

$$\sigma = E\varepsilon \tag{5.3}$$

The quantity E is called the *elastic modulus*, and it has units of force per unit area, such as GPa or Mpsi. The elastic modulus is a material property, and it is measured simply as the slope of the stress–strain curve in its straight-line region. By applying Equations 5.1 and 5.2, the amount that the rod stretches when it is loaded below the proportional limit becomes

$$\Delta L = \varepsilon L$$
$$= \left(\frac{\sigma}{E}\right) L$$
$$= \frac{F/A}{E} L$$
$$= \frac{FL}{EA} \tag{5.4}$$

or in terms of its stiffness

$$F = k\Delta L \qquad \text{where} \quad k = EA/L \tag{5.5}$$

A material's elastic modulus is related to the strength of its interatomic bonds, recognizing that metals and most other engineering materials are combinations of many chemical elements. As one example, steel alloys contain different weight fractions of such elements as carbon, molybdenum, manganese, chromium, and nickel. A material known as 1020 grade steel contains 0.18–0.23% carbon, 0.30–0.6% manganese, and a maximum of 0.04% phosphorus and 0.05% sulfur. As is the case for all steels, the alloy is comprised primarily of iron, and so the value of E for one alloy of steel is not much different than that for another. A similar situation exists for aluminum alloys, and as a general guideline, you can use the numerical values

$$E_{\text{steel}} \approx 210 \text{ GPa} \approx 30 \text{ Mpsi}$$

$$E_{\text{aluminum}} \approx 70 \text{ GPa} \approx 10 \text{ Mpsi}$$

The elastic modulus of aluminum is lower than that of steel by roughly a factor of 3. In practice, the elastic modulus for a specific sample could differ from these values, and so in critical applications, properties should be measured whenever possible. Property data for several generic metals are listed in Table 5.2.

As we described in Figure 5.2, after a rod is stretched in tension, it will also have a slightly smaller diameter. Conversely, the diameter would be slightly larger if a compressive force had been applied to it. This cross section effect is known as Poisson contraction or expansion, and it represents a dimensional change perpendicular to the applied force. When a soft material (such as a rubber band) is stretched, it is often possible to observe those dimensional changes visually and without any special equipment. For metals, the changes must be measured using precision laboratory instrumentation.

The material property that quantifies changes in the cross section is called the *Poisson ratio*, ν (the lowercase Greek character nu). It is defined in Figure 5.2 by the strains in the rod's diameter and length directions:

TABLE 5.2   Elastic and Weight Density Properties of Selected Metals*

| Material | Modulus of Elasticity, E | | Poisson's Ratio, ν | Weight Density | |
	Mpsi	GPa		lb/ft^3	kN/m^3
Aluminum alloy	10	72	0.32	172	27
Brass and bronze	16	110	0.33	536	84
Copper	17	121	0.33	552	86
Steel alloy	30	207	0.30	483	76
Stainless steel	28	190	0.30	483	76
Titanium alloy	16	114	0.33	276	43

* The numerical values given are representative, and values for specific materials could vary with composition and processing.

$$\nu = \frac{\varepsilon_{\text{diameter}}}{\varepsilon_{\text{length}}} = -\frac{\Delta d/d}{\Delta L/L} \tag{5.6}$$

In terms of the rod's elongation ΔL along its length, the change in diameter is given by

$$\Delta d = -\nu d \frac{\Delta L}{L} \tag{5.7}$$

The negative sign in these equations sets the sign convention that tension (with $\Delta L > 0$) causes the diameter to contract ($\Delta d < 0$), and compression causes it to expand. For many metals, $\nu \approx 0.3$, with numerical values for most samples and alloys falling in the range 0.25–0.35, as listed in Table 5.2.

EXAMPLE 5.2

For the 10-mm-diameter steel U-bolt in Example 5.1, determine the (a) strain, (b) change in length, and (c) change in diameter of the bolt's 300-mm straight section.

SOLUTION

(a) We previously found the stress to be $\sigma = 51$ MPa. Because the bolt is made of steel having $E = 210$ GPa, the strain is

$$\varepsilon = \frac{\sigma}{E} = \frac{51 \times 10^6 \text{ Pa}}{210 \times 10^9 \text{ Pa}} = 2.42 \times 10^{-4}$$

or 0.0242%, a dimensionless number.

(b) The elongation is

$$\Delta L = \varepsilon L$$

$$= (2.42 \times 10^{-4})(0.3 \text{ m})$$

$$= 7.26 \times 10^{-5} \text{ m}$$

$$= 72.6 \, \mu\text{m}$$

where we have used the SI prefix micro- (μ) to represent the factor of one-millionth. This extension is small indeed, approximately the same as the diameter of a human hair.

(c) The diameter change is even smaller and is found from the calculation

$$\Delta d = -\nu d \frac{\Delta L}{L}$$

$$= -(0.3)(0.01 \text{ m}) \frac{7.26 \times 10^{-5} \text{ m}}{0.3 \text{ m}}$$

$$= -7.26 \times 10^{-7} \text{ m}$$

$$= -726 \text{ nm}$$

where we have used $\nu = 0.3$ for generic steel. By comparison, the wavelength of light in a helium–neon laser is 632.8 nm. The bolt's diameter therefore contracts by an amount only somewhat more than 1 wavelength of light. ∎

Elastic limit $\dashrightarrow$ Returning to our discussion of the stress–strain behavior in Figure 5.7, point B is called the *elastic limit*. For loading between points A and B, the material continues to behave elastically, and it will spring back upon removal of the force, but the stress and strain are no longer proportional. As the load increases further beyond B, the material begins to show a permanent set. Yielding occurs in the region between B and C; for small changes in stress, the rod experiences a large

Yielding $\dashrightarrow$ change in strain. In the yielding region, the rod stretches substantially even as the force grows only slightly because of the shallow slope in the stress–strain curve. For that reason, the onset of yielding is often taken by engineers

Yield strength $\dashrightarrow$ as an indication of failure. The value of stress in region B–C defines a quantity called the *yield strength*, S_y. As the load is increased even further beyond

Ultimate strength $\dashrightarrow$ point C, σ grows to the ultimate tensile strength S_u at point D. That value represents the largest stress that the material is capable of sustaining. As the test continues, the stress in the figure actually decreases owing to a reduction in the rod's cross-sectional area, until the sample fractures at point E.

Table 5.3 lists yield and ultimate strength data for several metals. As for elastic modulus and Poisson ratio, the strength values for specific samples could differ from those listed in the table. Whenever possible, properties should be measured directly, or the material's supplier should be contacted. Table 5.3 includes property data for

TABLE 5.3   Yield and Ultimate Strength Properties of Selected Metals*

Material		Ultimate Strength, S_u		Yield Strength, S_y	
		ksi	MPa	ksi	MPa
Alloy steels	1020-HR	66	455	42	290
	1045-HR	92	638	60	414
	4340-HR	151	1041	132	910
Stainless steels	303-A	87	600	35	241
	316-A	84	579	42	290
	440C-A	110	759	70	483
Aluminum alloys	3003-A	16	110	6	41
	6061-A	18	124	8	55
	6061-T6	45	310	40	276
Copper alloys	Naval brass-A	54	376	17	117
	Cartridge brass-CR	76	524	63	434
Titanium alloy	Commercial	80	551	70	482

* The numerical values given are representative, and values for specific materials could vary with composition and processing. A = annealed, HR = hot-rolled, CR = cold-rolled, and T = tempered.

carbon and stainless steels, aluminum alloys, copper alloys, and a commercial titanium alloy. Grade 1020 steel, for instance, is a common medium-grade steel, while 4340 is a higher-strength but more expensive material. For stainless steels, the 316 alloy is sometimes used for fasteners and corrosion-resistant pipe fittings, and the hard material 440C is well suited for bearings and bushings. Aluminum alloy 3003 is often provided in sheet form, and it is easily bent and shaped to form boxes and covers for electronic components. The 6061 alloy is available either annealed (A) or tempered (T6), processing steps that involve heat treatment, quenching, and artificial aging. Those materials are frequently used for machined aluminum components. Copper alloys such as brasses (which are yellowish alloys of copper and zinc) and bronzes (brownish alloys of copper and tin) do not have particularly high strength, but they are corrosion resistant and easily joined by soldering. Copper alloys are used in such fluid-flow applications as tubing in condensers and heat exchangers. Titanium alloys offer a high strength-to-weight ratio and excellent corrosion resistance, but they are expensive and more difficult to machine than other materials. They are used in chemical industrial pipes, gas turbine blades, high-performance aircraft structures, submarines, and other demanding material applications.

Force–deflection and stress–strain curves are measured on a device called a *materials testing machine*. Figure 5.8 shows an example of equipment in which a computer both controls the test stand and records data. During tensile testing, a specimen such as a rod of steel is clamped by two jaws that are gradually pulled apart during the test, thus placing the specimen in tension. A sensor that measures force is attached to one of the jaws, and a second sensor, called an *extensometer*, is used to measure the amount that the sample stretches. A computer records the force and elongation data during the measurement, and those values are then converted to stress and strain.

Figure 5.9 shows a stress–strain curve that was measured for a sample of low-carbon steel. We can use that curve to determine the elastic modulus and yield strength of this particular alloy. The curve is nearly straight for small strain (up to about 0.2%), and the modulus E is determined from the slope in that region. The strain was zero

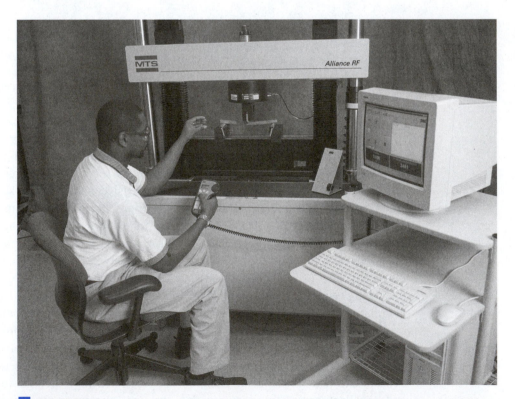

■ **FIGURE 5.8** A materials testing machine. The engineer is conducting a test in which a metal bar is bent between two supports, and the computer records the force and deflection data.
Source: Reprinted with permission of MTS Systems Corporation.

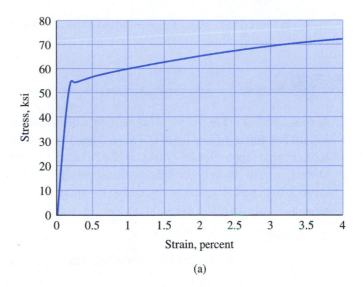

(a)

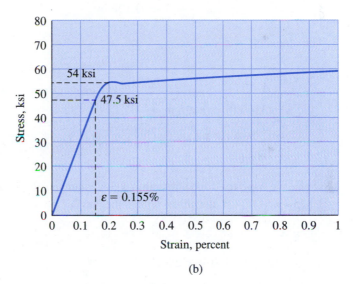

(b)

■ **FIGURE 5.9** Measured stress–strain curve for a steel alloy shown (a) over a wide range of strain, and (b) magnified to highlight the proportional and yielding regions.

when no stress was applied, and as read off the figure, the specimen was subjected to 47.5 ksi at a strain of 0.00155 (0.155%). Following Equation (5.3), the elastic modulus is

$$E = (47.5 \text{ ksi})/0.00155$$

$$= 3.06 \times 10^4 \text{ ksi}$$

$$= 30.6 \text{ Mpsi}$$

which is quite close to our rule-of-thumb value, 30 Mpsi. For this particular grade of steel, the yield point is also evident in Figure 5.9(b), and we measure $S_y \approx 54$ ksi.

EXAMPLE 5.3

A round rod is made from a steel sample with stress–strain characteristics shown in Figure 5.9. It has a diameter of 0.5 in. and length of 1 ft, and it is subjected to a 3500-lb tensile force (approximately the weight of a sedan automobile). Calculate (a) the stress and strain in the rod, (b) the amount that it stretches, (c) its change in diameter, (d) how close it is to yielding, and (e) its spring constant. (f) If the force was only 1000 lb, by what amount would the rod stretch?

SOLUTION

(a) This situation is depicted in Figure 5.2. The rod's cross-sectional area is

$$A = \frac{\pi d^2}{4} = \frac{\pi (0.5 \text{ in.})^2}{4} = 0.196 \text{ in}^2$$

The stress is

$$\sigma = \frac{F}{A}$$
$$= (3500 \text{ lb})/(0.196 \text{ in}^2)$$
$$= 1.79 \times 10^4 \text{ psi}$$
$$= 17.9 \text{ ksi}$$

where we have used the stress unit ksi to represent 1000 psi. With the conversion that 1 Mpsi = 1000 ksi, the strain within the rod is

$$\varepsilon = \frac{\sigma}{E}$$
$$= (17.9 \text{ ksi})/(30.6 \text{ Mpsi})$$
$$= (17.9 \text{ ksi})/(3.06 \times 10^4 \text{ ksi})$$
$$= 5.85 \times 10^{-4}$$

or 0.06%. We have expressed the elastic modulus in the units ksi to be compatible with the units for stress appearing in the numerator. Because stress and elastic modulus have the same units, they cancel, as expected, when calculating strain.

(b) The rod's elongation under the 3500-lb load is

$$\Delta L = \varepsilon L$$
$$= (5.85 \times 10^{-4})(12 \text{ in.})$$
$$= 0.007 \text{ in.}$$

By comparison, a sheet of writing paper is about 3- or 4-thousandths of an inch thick. The steel rod stretches by about the thickness of two sheets of standard paper, even under the car's seemingly large weight.

(c) Using $\nu = 0.3$ for steel, the strain across the rod's diameter is

$$\varepsilon_{\text{diameter}} = -\nu\varepsilon_{\text{length}}$$
$$= -(0.3)(5.85 \times 10^{-4})$$
$$= -1.76 \times 10^{-4}$$

The diameter therefore changes by

$$\Delta d = \varepsilon_{\text{diameter}}\, d$$
$$= (-1.76 \times 10^{-4})(0.5 \text{ in.})$$
$$= -8.77 \times 10^{-5} \text{ in.}$$

and the negative sign indicates that the diameter contracts. To place that small change in perspective, if the rod's initial diameter had been measured to five significant digits of accuracy as 0.50000 in., the diameter after extension would be 0.49991 in. Indeed, measuring a change at the fifth decimal place requires sensitive and well-calibrated instrumentation.

(d) Because the stress in the rod is 17.9 ksi, and the yield strength is 54 ksi, the rod has been loaded $(17.9 \text{ ksi})/(54 \text{ ksi}) = 0.33$ of the way to yielding.

(e) The spring constant is determined from its elastic modulus, cross-sectional area, and length according to

$$k = \frac{EA}{L}$$
$$= (30.6 \times 10^6 \text{ psi})(0.196 \text{ in}^2)/(12 \text{ in.})$$
$$= 4.99 \times 10^5 \text{ lb/in.}$$

(f) With a force of only 1000 lb, the rod's stretch would be reduced to

$$\Delta L = \frac{F}{k}$$
$$= (1000 \text{ lb})/(4.99 \times 10^5 \text{ lb/in.})$$
$$= 0.002 \text{ in.} \qquad \blacksquare$$

For aluminum and other nonferrous metals, the sharp corner and narrow yielding region B–C in Figure 5.7 is generally not seen. For such cases, a technique called the "0.2% offset method" is instead used to determine S_y. As an illustration, Figure 5.10 shows a stress–strain curve that was measured for a sample of aluminum. The modulus is again determined by drawing a straight line through the origin to fit the curve in the proportional region, as shown in Figure 5.10(b). For instance, when the stress is 15 ksi,

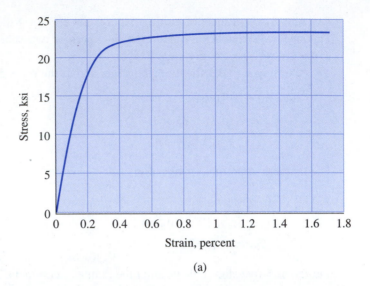

(a)

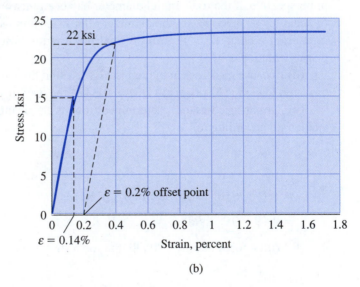

(b)

FIGURE 5.10   Measured stress–strain curve of an aluminum sample shown (a) over a wide range of strain and (b) magnified to highlight the 0.2% offset point.

the strain on the best-fit line is $\varepsilon = 0.0014$. The sample's modulus therefore becomes $E = (15\ \text{ksi})/(0.0014) \approx 1.1 \times 10^4\ \text{ksi} \approx 11\ \text{Mpsi}$. However, unlike the steel specimen in Figure 5.9, the onset of yielding is not obvious, and the 0.2% offset method is instead used to determine S_y. The yield stress is defined by the curve's intersection point with a line drawn at slope E, but offset from the

0.2% offset method

origin by $\varepsilon_{\text{offset}} = 0.002 = 0.2\%$. In Figure 5.10(b), the straight construction line drawn with slope 11 Mpsi from the offset point intersects the stress–strain curve at $S_y = 22$ ksi. That value is taken as the level where the sample begins to yield appreciably.

EXAMPLE 5.4

A 50-cm-long rod has a rectangular cross section with dimensions 5 by 10 mm, and it is made from the aluminum alloy with stress–strain characteristics depicted in Figure 5.10. (a) By what length must the rod be stretched in order that its stress reaches the 0.2% yield point? (b) What force is carried by the rod at that point?

SOLUTION

(a) The rod's stress at the yield point is $\sigma = S_y$. Solving for the extension, we have

$$\Delta L = \varepsilon L$$

$$= \left(\frac{\sigma}{E}\right)L$$

$$= \left(\frac{S_y}{E}\right)L$$

in which the dimensions of the rod's cross section do not appear. The rod could have twice as large a cross section, or be twice as small, and the amount that it stretches at the yield point would be unchanged. Becuase both S_y and E are known in the USCS, but the rod's length is known in the SI, we must convert the units using

$$E = (11 \text{ Mpsi})\left(6.895 \ \frac{\text{GPa}}{\text{Mpsi}}\right) = 75.8 \text{ GPa}$$

$$S_y = (22 \text{ ksi})\left(6.895 \ \frac{\text{MPa}}{\text{ksi}}\right) = 152 \text{ MPa}$$

The rod begins to yield when it is stretched by

$$\Delta L = \frac{152 \text{ MPa}}{75.8 \text{ GPa}}(0.5 \text{ m}) = \frac{152 \text{ MPa}}{75,800 \text{ MPa}}(0.5 \text{ m}) = 0.001 \text{ m} = 1 \text{ mm}$$

(b) As the rod begins to yield, the force carried by it is

$$F = \sigma A$$

$$= S_y A$$

$$= (152 \times 10^6 \text{ N/m}^2)(0.005 \text{ m})(0.01 \text{ m})$$

$$= 7.6 \times 10^3 \text{ N}$$

which, of course, does depend on the cross-sectional dimensions. ∎

5.4 SHEAR

The tensile stress σ in Figure 5.3(c)–(d) is directed along the rod's length, and it is perpendicular to the cross section. Another type of stress, called *shear stress*, is also exposed when one imagines making a slice through a structure's cross section. However, shear stress is oriented in the same plane as the cross section, that is, in the direction parallel to the surface of the slice. Roughly speaking, tension is associated with a component that is being stretched or pulled apart, whereas shear stress occurs when something is being sliced or cut. Thus, the two types of stress differ in the sense that they act in perpendicular directions. Like tensile stress, the units for shear stress remain force-per-unit-area.

Shear plane

- - - - - - - - - - - - - - →

Shear force

- - - - - - - - - - - - - - →

Consider the block of material in Figure 5.11 that is being pressed downward and forced between two rigid supports. As the force F is applied, the material tends to be sliced, sheared, or cut along the two edges marked as shear planes in Figure 5.11(a). In Figure 5.11(b), a free body diagram of the block is shown, and equilibrium in the vertical direction requires that $V = F/2$. The two forces V are called *shear forces*, and they are oriented parallel to the shear planes. The shear forces are, in fact, distributed within the material over the entire cross section so that V is the resultant of the shear stress shown in Figure 5.11(c). Analogous to the definition of tensile stress, the shear stress, τ, is defined

$$\tau = \frac{V}{A} \tag{5.8}$$

where A is the area of the cross section.

Two connections that are seen in practice involve single and double shear. The terminology refers to the manner in which shear forces are transmitted between two components that are joined together. Figure 5.12 shows these configurations for

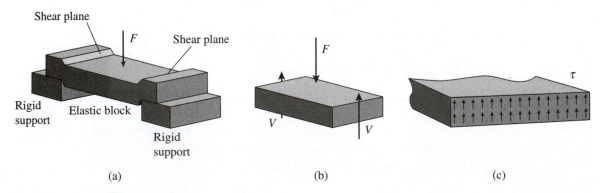

(a)   (b)   (c)

■ FIGURE 5.11   Shear forces V and shear stresses τ act on an elastic block that is being pressed between two rigid supports.

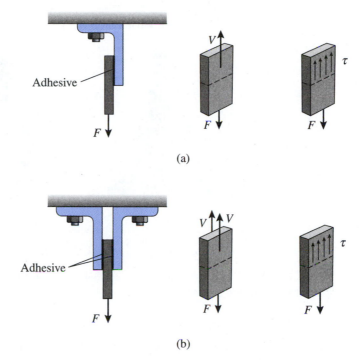

(a)

(b)

■ **FIGURE 5.12**   Adhesively bonded connections that are placed in (a) single shear with $V = F$ and (b) double shear with $V = F/2$. The adhesive layers are indicated by the heavily weighted lines.

Single shear
- - - - - - - - - - - - - - - ➤

Double shear
- - - - - - - - - - - - - - - ➤

the illustrative case of adhesively bonded lap joints. By using free body diagrams, we imagine disassembling the pieces to expose the shear forces V acting within the adhesive layers. For single shear in Figure 5.12(a), the full load is carried by the adhesive, and $V = F$. Double shear is illustrated in Figure 5.12(b). Because two surfaces now share the load, $V = F/2$ and the shear stress is halved. Similar situations occur when components are connected by bolts, rivets, and welds, as we discuss in the following example.

EXAMPLE 5.5

The hinged connection in Figure 5.13(a) is made by the $\frac{3}{8}$-in.-diameter clevis pin. The threaded rod is subjected to 350 lb of tension, and that force is transferred to the fixed base. Determine the shear stress present in the pin.

SOLUTION

The bolts attached to the rigid mounting surface are loaded along their length in tension, but the pin is loaded perpendicular to its length in shear. Referring to the free body

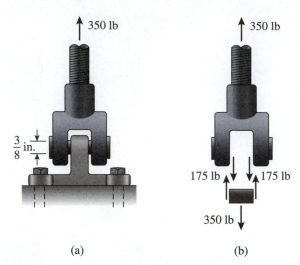

(a)   (b)

■ **FIGURE 5.13**   (a) Clevis pin that is loaded in double shear in Example 5.5. (b) Free body diagram of the connection.

diagrams in Figure 5.13(b), the pin is subjected to double shear, and $V = 175$ lb. The shear stress in the pin is therefore given by

$$\tau = (175 \text{ lb})/(\pi(0.375 \text{ in.})^2/4)$$

$$= 1.59 \times 10^3 \text{ psi}$$

$$= 1.59 \text{ ksi}$$

EXAMPLE 5.6

In Example 4.6, we found that the wire cutter's hinge pin B must support a force of magnitude 385 N. If the pin is loaded in single shear and its diameter is 7 mm, determine the shear stress present.

SOLUTION

The force transmitted by pin B from one combined jaw/handle to the other was found previously by using the requirements of static equilibrium. We now extend that analysis to examine how intensely the material of the pin is being loaded. The pin's cross-sectional area is

$$A = \frac{\pi(0.007 \text{ m})^2}{4} = 3.85 \times 10^{-5} \text{ m}^2$$

and the shear stress becomes

$$\tau = \frac{V}{A}$$

$$= (385 \text{ N})/(3.85 \times 10^{-5} \text{ m}^2)$$

$$= 10.0 \times 10^6 \text{ Pa}$$

$$= 10 \text{ MPa}$$

■

EXAMPLE 5.7

The spur gear shown in Figure 5.14 is attached to the 1-in.-diameter shaft by a key having $\frac{1}{4}$-in. by $\frac{1}{4}$-in. cross section and length $1\frac{3}{4}$ in. The key transmits torque by fitting into matching slots in the shaft and gear. When the 1500-lb force acts on the gear as shown, determine the shear stress in the key along plane $A–A$.

SOLUTION

Torque is transmitted between the gear and the shaft through the key, and plane $A–A$ is the shear plane along which the key tends to be sliced. The shear force is determined by applying the condition of rotational equilibrium for the gear. Taking moments about the gear's center, the shear force acting on the key is

$$V = \frac{(1500 \text{ lb})(2.5 \text{ in.})}{0.5 \text{ in.}} = 7500 \text{ lb}$$

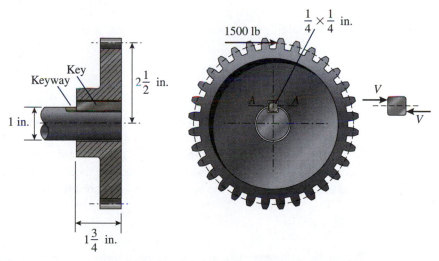

FIGURE 5.14   Determining the shear stress in the gear key in Example 5.7.

This force is distributed over the cross section of the key having area

$$A = (1.75 \text{ in.})(0.25 \text{ in.}) = 0.438 \text{ in}^2$$

The shear stress then becomes

$$\tau = \frac{V}{A} = \frac{7500 \text{ lb}}{0.438 \text{ in}^2} = 1.71 \times 10^4 \text{ psi} = 17.1 \text{ ksi} \qquad \blacksquare$$

5.5 FACTOR OF SAFETY

Mechanical engineers determine the shape, dimensions, and materials for a wide range of hardware. The analyses that support those design decisions take into account the tensile, compressive, and shear stresses that are present. Designers are aware that a component can break or otherwise be rendered useless through a variety of mechanisms. It could yield and take on a permanently deformed shape, fracture suddenly into many pieces, or be damaged through corrosion. Mechanical engineers perform the calculations and tests that are necessary to ensure reliability against those failure modes and other possible ones. Engineers rely on analysis, experiments, judgment, and design codes when they make decisions of that nature.

In this section, we discuss failure analysis for the prototypical case of yielding in either tension or shear. Such an analysis predicts the onset of yielding in ductile materials and is useful to prevent the material from being loaded to, or above, its yield stress. Yielding is only one of many possible failure mechanisms, and our analysis in this section will not offer predictions regarding other kinds of failure.

When a straight rod is placed in tension as in Figure 5.2, the possibility of it yielding is assessed by comparing the stress, σ, to the material's yield strength, S_y. Failure due to ductile yielding is predicted if $\sigma \geq S_y$. Engineers define the tensile factor of safety as

$$n_{\text{tension}} = \frac{S_y}{\sigma} \qquad (5.9)$$

If $n > 1$, this viewpoint predicts that the component will not yield, and if $n < 1$, failure is expected to occur. For components that are loaded in pure shear, on the other hand, the shear stress τ is instead compared to the shear yield strength, S_{sy}. As developed in more advanced treatments of stress analysis, one theory for material failure relates the yield strength in shear to the value in tension according to

$$S_{sy} = \frac{S_y}{2} \qquad (5.10)$$

The shear yield strength can therefore be determined from S_y values that are obtained from standard tensile testing, as in Table 5.3. To evaluate the possibility of a ductile material yielding in shear, we compare stress and strength according to the shear factor of safety

$$n_\text{shear} = \frac{S_{sy}}{\tau} \tag{5.11}$$

The value of n that an engineer chooses for a particular design will depend on many parameters, including the designer's background, experience with components similar to the one being analyzed, the amount of testing that will be done, the material's reliability, the consequences of failure, maintenance and inspection procedures, and cost. Certain spacecraft components might be designed with n only slightly greater than 1 in order to reduce weight, which is at a premium in aerospace applications. On the other hand, to offset that seemingly small safety factor, those components will be extensively analyzed and tested, and they will be developed by engineers having a great deal of collective experience. When forces and load conditions are not known with certainty, or when the consequences of a component's failure would be significant or endanger life, large values of n are appropriate. Engineering handbooks and design codes often recommend ranges for safety factors, and those references should be used whenever possible.

SUMMARY

One of the primary functions of mechanical engineers is to design structures and machines so that they do not break. Analyses of stress, strain, and strength are tools used by engineers to determine if a component will be overloaded or if it will deform excessively. Generally, engineers conduct a stress analysis during the design process, and the results of those calculations are used to guide the choice of materials and dimensions. As we introduced in Chapter 1, when the geometry of a component or the loading circumstances are particularly complicated, engineers use computer-aided tools to calculate stresses and deflections. Figure 5.15 illustrates the results of two such computer-aided stress analyses for a clevis connection (as described in Example 5.5) and a pillow block bearing (as discussed in Section 3.2).

In this chapter, we have discussed the loading conditions known as tension, compression, and shear, and a failure mechanism called ductile yielding. We further introduced the following concepts:

Stress is the intensity of a force distributed over an area of material. Depending on the direction in which the stress acts, it can be tension or shear.

Strain is defined as the change in length per unit of original length. Because of its definition as a ratio of two lengths, strain is a dimensionless quantity, and it is often expressed as a decimal percent. At a strain of 0.1%, a bar that was 1 meter long will have stretched by a factor of 0.001, or 1 millimeter.

Strength describes the ability of a material to support the stresses acting on it. Some measures of strength include the yield strength, S_y, and the ultimate strength, S_u.

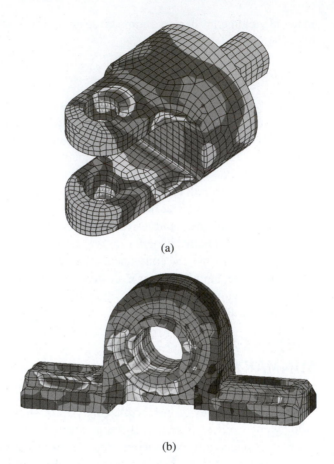

(a)

(b)

█ FIGURE 5.15 Computer-aided stress analyses of (a) a clevis connection and (b) a pillow block bearing.
Source: Reprinted with permission of Algor Corporation.

TABLE 5.4 Quantities, Symbols, and Units That Arise in the Analysis of Stresses and Material Properties

		Conventional Units	
Quantity	**Conventional Symbols**	**USCS**	**SI**
Tensile stress	σ	psi, ksi, Mpsi	Pa, kPa, MPa
Shear stress	τ	psi, ksi, Mpsi	Pa, kPa, MPa
Elastic modulus	E	Mpsi	GPa
Yield strength	S_y	ksi	MPa
Ultimate strength	S_u	ksi	MPa

TABLE 5.4 (*continued*)

Quantity	Conventional Symbols	Conventional Units	
		USCS	SI
Shear yield strength	S_{sy}	ksi	MPa
Strain	ε	—	—
Poisson ratio	ν	—	—
Factor of safety	n	—	—
Spring constant	k	lb/in.	N/m

The important quantities introduced in this chapter, common symbols for them, and their units are summarized in Table 5.4.

SELF-STUDY AND REVIEW

1. Sketch stress–strain curves for steel and aluminum.
2. What is the difference between elastic and plastic behavior?
3. Define the following terms: *elastic modulus*, *proportional limit*, *elastic limit*, *yield point*, and *ultimate point*.
4. What are the approximate numerical values for the elastic modulus of steel and aluminum?
5. How is the yield strength found using the 0.2% offset method?
6. If a rod is stretched in tension, how does Poisson's ratio, ν, relate the change in diameter to the change in length?
7. Discuss some of the differences in alloys, material properties, and applications for steel, aluminum, and other metals.
8. In what ways do tensile and shear stresses differ?
9. How is the shear yield strength S_{sy} related to the yield strength S_y obtained from a tensile test?
10. What is meant by the term *factor of safety*?
11. Discuss the trade-off between a factor of safety being too large or too small.

PROBLEMS

1. Find a real physical example of a mechanical structure or machine that has tensile stress present. (a) Make a clear, labeled drawing of the situation. (b) Estimate the dimensions of the structure or machine and the magnitudes and directions of the forces that act on it. Show these on the drawing. Briefly explain why you estimate the dimensions and forces to have the numerical values that you assigned. (c) Calculate the magnitude of the stress.

2. A steel cable of diameter $\frac{3}{16}$ in. is attached to an eyebolt and tensioned to 500 lb. Calculate the stress in the cable, and express it in the units psi, ksi, Pa, kPa, and MPa.

500 lb

3. When a 120-lb woman stands on a snow-covered trail, she sinks slightly into the snow because the compressive stress between her ski boots and the snow is larger than the snow can support without crumbling. Her cross-country skis are 65 in. long and $1\frac{7}{8}$ in. wide. After estimating the dimensions of a boot's footprint, calculate the percent reduction in stress applied to the snow when she is wearing skis instead of boots.

4. As a machinist presses the handles of the compound action bolt cutters, link AB carries a 7.5-kN force, as shown. If the link has a 14-mm by 4-mm rectangular cross section, calculate the tensile stress within it.

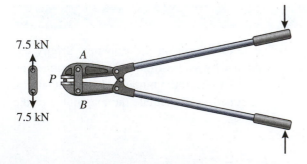

7.5 kN

7.5 kN

5. (a) By using either the vector algebra or polygon methods for finding a resultant, determine the magnitude of F that will cause the net effect of the three forces in the diagram to act vertically. (b) For that value of F, determine the stress in the bolt's $\frac{3}{8}$ in.-diameter straight shank.

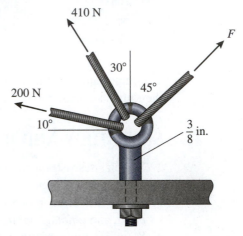

410 N

200 N

6. The band saw blade in a machine shop cuts through a work piece that is fed between the two guide blocks B in the diagram. To what

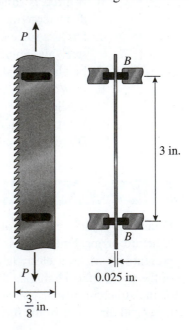

tension P should the blade be set if its stress should be 5 ksi during use? Neglect the small size of the teeth relative to the blade's width.

7. An aluminum rod with a diameter of 8 mm and an original length of 30 cm has lines scribed on it that are exactly 10 cm apart. After a 2.11-kN force is applied, the separation between the lines is measured to be 10.006 cm. (a) Calculate the stress and strain in the rod. (b) To what total length has the rod stretched?

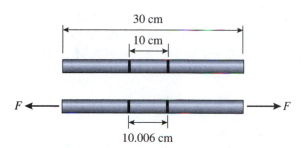

8. The tires of a sport utility wagon are $6\frac{1}{2}$ in. wide and deform slightly under the vehicle's 4555-lb weight. Each tire contacts the ground over a distance of $4\frac{1}{4}$ in. measured along the vehicle's length. Calculate the compressive stress between each tire and the road. The wheelbase dimensions are shown in the figure.

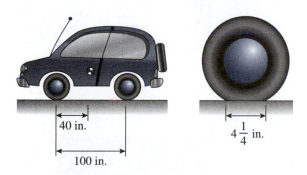

9. Determine the elastic modulus and the 0.2% offset yield strength for the plastic sample

with the stress–strain curve shown in the figure.

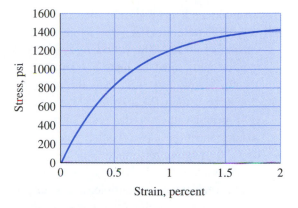

10. A 1-ft-long rod is made from the plastic material of Problem 9. By what amount must the rod be stretched from its original length in order to begin yielding?

11. The steel bolt and anchor assembly in the figure is used to reinforce the roof of a passageway in an underground coal mine. The $d = \frac{5}{8}$-in.-diameter bolt is $L = 34$ in. long. In installation, the bolt is tensioned to 5000 lb. Calculate the stress, strain, and extension of the bolt if it is formed from a 1045 steel alloy.

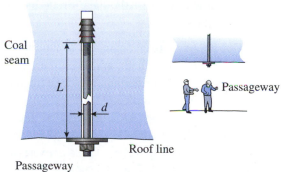

12. A circular rod of length 25 cm and diameter 8 mm is made of grade 1045 steel. (a) Calculate the stress and strain in the rod, and its extension, when it is subjected to 5 kN of tension. (b) At what force would the rod begin

to yield? (c) By what amount would the rod have to be stretched beyond its original length in order to yield?

13. An engineer determines that a 40-cm-long rod of 1020 grade steel will be subjected to a tension of 20 kN. The following two design requirements must be met: The stress must remain below 145 MPa, and the rod must stretch less than 0.125 mm. Determine an appropriate value for the rod's diameter in order to meet these two requirements. Round up to the nearest millimeter when reporting your answer.

14. Find a real physical example of a mechanical structure or machine that has shear stress present. (a) Make a clear, labeled drawing of the situation. (b) Estimate the dimensions of the structure or machine and the magnitudes and directions of the forces that act on it. Show these on the drawing. Briefly explain why you estimate the dimensions and forces to have the numerical values that you assigned. (c) Calculate the magnitude of the stress.

15. Adhesive tape is capable of supporting relatively large shear stress, but it is not able to support significant tensile stress. In this problem, you will measure the shear strength of a piece of tape.

 • Cut about a dozen segments of tape having identical length L and width b. The exact length isn't important, but the segments should be easily handled.
 • Develop a means to apply and measure the pull force F on the tape. Use, for instance, dead weights (cans of soda or exercise weights) or a small fishing scale.
 • Attach a segment of tape to the edge of a table, with only a portion of tape adhering to the surface. In your tests, consider lengths ranging between a fraction of an inch and several inches.

 • Being careful to apply the pull force straight along the tape, measure the value F necessary to cause the adhesive layer to slide or shear off the table. Tabulate pull force data for a half-dozen different lengths a.
 • Make graphs of pull force and shear stress versus a. From the data, estimate the value of the shear stress above which the tape will slide and come loose from the table.
 • At what length a did the tape break before it sheared off the table?
 • Repeat the tests for the orientation in which F is applied perpendicular to the surface, tending to peel the tape instead of shearing it. Compare the tape's strengths for shear and peeling.

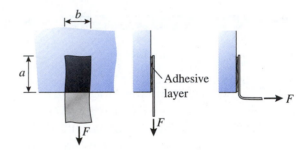

16. The small steel plate in the diagram is connected to the right angle bracket by a 10-mm-

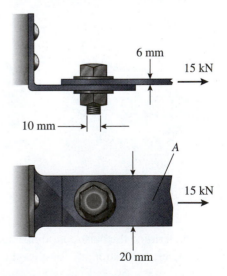

diameter bolt. Determine the tensile stress at point *A* in the bar and the shear stress in the bolt.

17. A 600-lb force acts on the vertical plate in the diagram, which in turn is connected to the horizontal truss by five $\frac{3}{16}$-in.-diameter rivets. (a) If the rivets share the load equally, determine the shear stress in them. (b) In a worst-case scenario, four of the rivets have corroded and the load is carried by only one rivet. What is the shear stress in this case?

600 lb

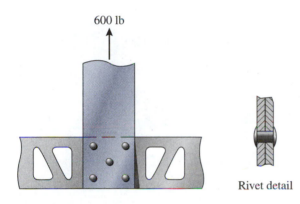

Rivet detail

18. The accompanying figure shows detail of the connection *B* in the concrete trough from Problem 17 in Chapter 4. Determine the shear stress acting in the shackle's $\frac{3}{8}$-in. bolt.

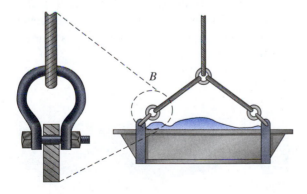

19. The tie-down mount in the figure is bolted to the deck of a cargo truck and restrains the steel

cable under 1.2 kN of tension. Determine the shear stress in the 6-mm-diameter bolt. Neglect the small horizontal and vertical offset distances between the bolt and the cable.

1.2 kN

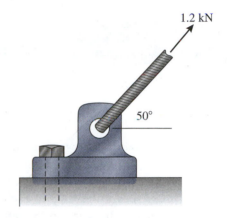

50°

20. A spur gear transmits 35 N·m of torque to the 20-mm-diameter driveshaft in the accompanying figure. The 5-mm-diameter setscrew threads onto the gear's hub, and it is received in a small hole that is machined in the shaft. Determine the shear stress in the screw along the shear plane *B*–*B*.

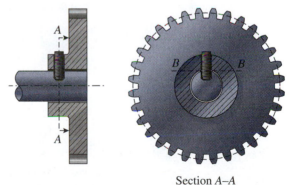

Section *A–A*

21. The accompanying figure shows the closed end of a plastic pipe that carries deionized water in a microelectronics clean room. The water pressure is $p_0 = 50$ psi, and the cap is attached to the end of the pipe by an adhesive. Calculate the shear stress τ present in the adhesive.

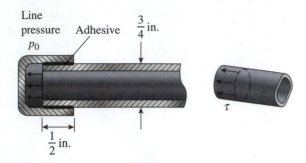

Line pressure p_0 Adhesive $\frac{3}{4}$ in.

$\frac{1}{2}$ in.

τ

22. A small stepladder has vertical rails and horizontal steps formed from C-section aluminum channel. Two rivets, one in front and one in back, secure the ends of each step. The rivets attach the steps to the left-hand and right-hand rails. A 200-lb person stands in the center of a step. If the rivets are formed of 6061-T6 aluminum, what should be the diameter d of the rivets? Use a factor of safety of 6, and round your answer to the nearest $\frac{1}{16}$ of an inch.

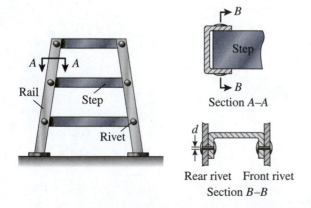

B

Step

B

Section A–A

A   A

Rail   Step

Rivet

d

Rear rivet   Front rivet

Section B–B

23. For the exercise of Problem 22, and for the most conservative design, at what location on the step should you specify in your calculation that the 200-lb person stand? Determine the rivet diameter for that loading condition.

24. A $\frac{3}{8}$-in. diameter bolt connects the marine propeller to a boat's $1\frac{1}{4}$-in. driveshaft. In order to protect the engine and transmission should the propeller accidentally strike an underwater obstacle, the bolt is designed to be cut when the shear stress acting on it reaches 25 ksi. Determine the contact force between the blade and obstacle that will cause the bolt to be sheared, assuming a 4-in. effective radius between the point of contact on the blade and the center of the driveshaft.

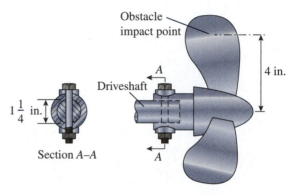

Obstacle impact point

4 in.

A

Driveshaft

$1\frac{1}{4}$ in.

Section A–A

A

25. A machinist squeezes the handle of visegrip pliers while loosening a frozen bolt. The connection at point A, which is shown in a magnified cross-sectional view, supports a 4.1-kN force. Determine (a) the shear stress in the 6-mm-diameter rivet at A and (b) the factor of safety against yielding if the rivet is formed from 4340 steel alloy.

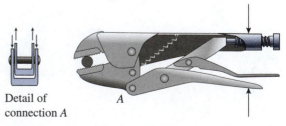

Detail of connection A

A

26. Compound lever shears cut through a piece of wire at A in the accompanying diagram. (a) By using the free body diagram of handle CD, determine the magnitude of the force at rivet D. (b) Referring to the magnified cross-sectional drawing of the connection at D, determine the shear stress in the rivet. (c) If the rivet is formed from 4340 steel alloy, what is the factor of safety?

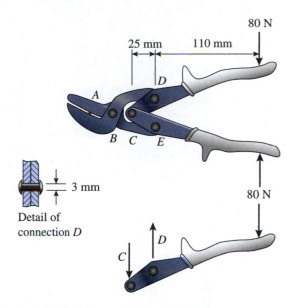

80 N

25 mm   110 mm

D

A

B C   E

3 mm

Detail of
connection D

80 N

D

C

6 Thermal and Energy Systems

6.1 OVERVIEW

Up to this point, we have explored the first three "elements" of the mechanical engineering profession: machine components and tools, forces in structures and fluids, and materials and stresses. In Chapter 1, mechanical engineers were described as developing equipment that either consumes or produces power. We therefore turn our attention in this chapter to the fourth element of mechanical engineering and the practical topics of thermal and energy systems. This field encompasses such hardware as internal combustion engines, aircraft propulsion, and electrical power generation through solar, hydroelectric, fossil fuel, and nuclear technologies. In the first portion of this chapter, we introduce the physical principles and terminology needed to understand the operation and efficiency of energy systems. Later, in Sections 6.6–6.8, we will apply those ideas in three case studies of common mechanical engineering equipment.

The characteristics of energy, the different forms that energy takes, and the methods by which energy can be converted from one form to another lie at the very heart of mechanical engineering. Of particular interest from the standpoint of designing equipment to consume or produce power are the processes for (1) converting chemical energy into heat, (2) converting heat into mechanical energy, and (3) moving energy between locations. In an automobile engine, for instance, gasoline is burned to release stored chemical energy as heat. The engine, in turn, converts that heat into mechanical energy—rotation of the engine's crankshaft—and ultimately, the automobile's motion (Figure 6.1). We will develop such concepts and apply them to a range of engineering hardware in the following sections. After completing this chapter, you should be able to:

- Convert numerical values for mechanical energy, heat, and work between the SI and USCS.
- Describe how heat is transferred by the processes of conduction, convection, and radiation.
- Discuss energy conservation for a mechanical system or a flowing fluid.

180

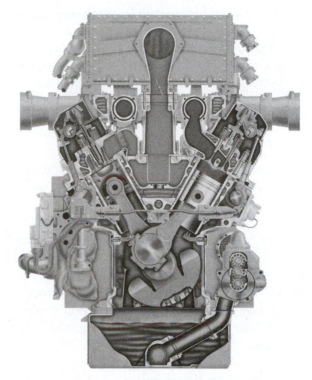

- Describe how heat engines operate, and understand the difference between real and ideal efficiencies.
- Explain how four-stroke and two-stroke internal combustion engines, electric power plants, and jet engines operate.
- Calculate the efficiency, power consumption, and power generation characteristics of certain machines encountered in mechanical engineering.

6.2 MECHANICAL ENERGY, WORK, AND POWER

Although you cannot see energy or hold it in your hand, it is needed to stretch an object, increase its velocity, raise its temperature, or elevate it. Energy can be stored in the form of gravitational potential energy as an object is lifted upward against gravity or as elastic potential energy when a spring is stretched.

Gravitational potential energy is associated with changing the elevation of an object within a gravitational field. Near the surface of the Earth, the acceleration of gravity is reasonably constant, and the values $g = 9.81$ m/s^2 $= 32.2$ ft/s^2 are usually accurate enough for most engineering calculations. The local value of g, however, does vary slightly with position on the Earth, and accelerations at the equator and poles differ by several percent. Gravitational potential energy is measured relative to a reference height (for instance, the ground or the top of a workbench), and it is stored by virtue of vertical position. The change ΔU in energy as an object moves through the vertical distance Δh is

$$\Delta U = mg\,\Delta h \tag{6.1}$$

where m is the object's mass. The potential energy increases ($\Delta U > 0$) when the object moves higher ($\Delta h > 0$). Conversely, the gravitational potential energy decreases ($\Delta U < 0$) when the object is lowered ($\Delta h < 0$).

For calculations involving energy and work, it is conventional to use the units joule (1 J $= 1$ N·m) in the SI, and ft·lb in the USCS. As described in Chapter 2, a prefix can be used in the SI to represent either small or large quantities; for instance, 1 kJ $= 10^3$ J and 1 MJ $= 10^6$ J. In some circumstances, the units of Btu (British thermal unit) and kW·h (kilowatt hour) are convenient, and we will discuss those choices later in this chapter. The conversion factors between units for energy and work are listed in Table 6.1. Reading off the first row, for instance, we see that the unit ft·lb is converted to the unit of J through the equivalence 1 ft·lb $= 1.356$ J, and from the last row, that 1 kW·h $= 3.6 \times 10^6$ J $= 3.6$ MJ.

Elastic potential energy is stored by an object when it is stretched or bent as described by Hooke's law (Section 5.3). For a spring having stiffness k, the elastic potential energy is

$$\Delta U = \frac{1}{2}k\Delta L^2 \tag{6.2}$$

where ΔL is the distance that the spring has been stretched or compressed. If the spring has original length L_0, and if it is stretched to the new length L after a force is applied, the elongation is $\Delta L = L - L_0$. Although ΔL can be positive (when the spring is stretched) or negative (compressed), the elastic potential energy is always positive. As

TABLE 6.1 Conversion Factors Between Various Units for Energy and Work in the USCS and SI

ft·lb	J	Btu	kW·h
1	1.356	1.285×10^{-3}	3.766×10^{-7}
0.7376	1	9.478×10^{-4}	2.778×10^{-7}
778.2	1055	1	2.930×10^{-4}
2.655×10^6	3.600×10^6	3413	1

discussed in Chapter 5, k has the units of force-per-unit length, and we generally use N/m and lb/in. in the SI and USCS, respectively, for stiffness. You should note that Equation 6.2 can be applied whether or not a machine component actually "looks" like a coiled spring. In Chapter 5, for instance, the force–deflection behavior of a rod in tension or compression was discussed, and in that case, the spring constant was given by Equation 5.5.

Kinetic energy is associated with an object's motion. As forces or moments act on a machine, they can cause it to move and store kinetic energy. The motion can be in the form of vibration (for instance, the cone of a stereo speaker), rotation (the flywheel on an engine), or translation (the straight-line motion of the piston in a compressor). For an object of mass m that moves in a straight line with velocity v, the kinetic energy is defined as

$$\Delta U = \frac{1}{2}mv^2 \tag{6.3}$$

You can verify that J and ft·lb are appropriate units for kinetic energy, just as they are for gravitational and elastic potential energy.

Work of a force is illustrated in Figure 6.2 by a piston that slides horizontally in its cylinder, as would be the case in either an automobile engine or an air compressor. In Figure 6.2(a), the piston moves to the right, and the force F is applied to compress the gas within the cylinder. On the other hand, if the gas is already at a high pressure and the piston moves to the left (Figure 6.2(b)), the force F resists that expansion. Those two situations might correspond to the compression and power strokes occurring in an automobile engine, for instance. The work ΔW of the force as the piston moves through the distance Δd is given by

$$\Delta W = F\Delta d \tag{6.4}$$

In Figure 6.2(a), the work is positive ($\Delta d > 0$) because the force acts in the same direction as the piston's motion. Conversely, the work is negative if the force opposes motion, as in Figure 6.2(b) with $\Delta d < 0$.

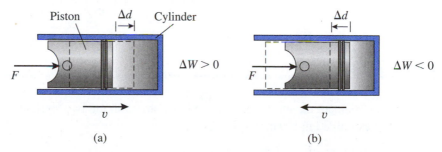

◼ **FIGURE 6.2**   Work of a force applied to a piston that slides within the cylinder of an engine or compressor. (a) $\Delta d > 0$: The force reinforces displacement as the gas in the cylinder is compressed. (b) $\Delta d < 0$: The force opposes displacement as the gas expands.

Power, the last of the quantities introduced in this section, is defined as the rate at which work is performed. When a force performs work during the interval of time Δt, the average power is defined

$$P_{\text{avg}} = \frac{\Delta W}{\Delta t} \tag{6.5}$$

As work is performed more rapidly, Δt becomes smaller, and the average power likewise increases. Engineers conventionally express power in the units of watt (1 W = 1 J/s) in the SI, and either ft·lb/s or horsepower (hp) in the USCS. Table 6.2 lists conversion factors between those choices of units. Reading off the third row, for instance, we see that horsepower is converted to ft·lb/s according to 1 hp = 550 ft·lb/s.

TABLE 6.2   Conversion Factors Between Various Units for Power in the USCS and SI

ft·lb/s	W	hp
1	1.356	1.818×10^{-3}
0.7376	1	1.341×10^{-3}
550	745.7	1

EXAMPLE 6.1

Beginning with the definitions of ft·lb/s and W, determine the conversion factor between them in Table 6.2.

SOLUTION

To relate those two units for power, we convert forces and lengths as follows:

$$1 \text{ ft·lb/s} = \left(1 \frac{\text{ft·lb}}{\text{s}}\right)\left(0.3048 \frac{\text{m}}{\text{ft}}\right)\left(4.448 \frac{\text{N}}{\text{lb}}\right) = 1.356 \text{ N·m/s} = 1.356 \text{ W} \qquad \blacksquare$$

EXAMPLE 6.2

Calculate the elastic potential energy stored in the two straight sections of the U-bolt examined in Examples 5.1 and 5.2.

SOLUTION

With the dimensions and material properties used in those examples, the spring constant for one straight section is found from Equation 5.5:

$$k = \frac{(210 \times 10^9 \text{ Pa})(7.85 \times 10^{-5} \text{ m}^2)}{0.3 \text{ m}} = 5.5 \times 10^7 \text{ N/m}$$

In Example 5.2, each straight section of the U-bolt was found to stretch by 72.6 μm. By using Equation 6.2, the elastic potential energy becomes

$$\Delta U = \frac{1}{2}(5.5 \times 10^7 \text{ N/m})(7.26 \times 10^{-5} \text{ m})^2 = 0.145 \text{ N·m} = 0.145 \text{ J}$$

Because the U-bolt has two identical straight sections, the stored energy is $2(0.145 \text{ J}) = 0.29$ J, a relatively modest amount of energy. ∎

EXAMPLE 6.3

In September of 2001, terrorists used commercial jetliners to destroy the twin towers of the World Trade Center in New York City. Calculate the kinetic energy of a Boeing 767 that is loaded to its maximum weight of 350,000 lb and travels at 400 mph.

SOLUTION

To apply Equation 6.3, we must first calculate the aircraft's mass

$$m = \frac{3.5 \times 10^5 \text{ lb}}{32.2 \text{ ft/s}^2} = 1.087 \times 10^4 \text{ slugs}$$

and convert speed into the consistent units of ft/s:

$$v = (400 \text{ mph})(5280 \text{ ft/mi})(2.778 \times 10^{-4} \text{ h/s}) = 587 \text{ ft/s}$$

The kinetic energy becomes

$$\Delta U = \frac{1}{2}(1.087 \times 10^4 \text{ slugs})(587 \text{ ft/s})^2 = 1.87 \times 10^9 \text{ ft·lb}$$

Referring to Table 6.1, that amount of energy is equivalent to

$$\Delta U = (1.87 \times 10^9 \text{ ft·lb})(1.356 \text{ J/ft·lb}) = 2.54 \times 10^9 \text{ J} = 2.54 \text{ GJ}$$

in the SI. To place that quantity of energy into perspective, we recall from Problem 18 in Chapter 2 that the unit "kiloton" is sometimes used to describe the energy produced by large explosions. Being approximately equivalent to the explosive energy of 1000 tons of high explosive, the kiloton is defined as 4.186×10^{12} J. With that conversion factor, we see that the aircraft's kinetic energy is comparable to 6×10^{-4} kiloton, or some 1200 lb of high explosives. ∎

EXAMPLE 6.4

Make an order-of-magnitude calculation to estimate the capacity, in the units of horsepower, of an electric motor that will power an elevator in a four-story office building.

The elevator car weighs 500 lb, and it will carry up to an additional 2500 lb of passengers and freight.

SOLUTION

We apply Equation 6.5 to estimate the motor's power rating. We estimate that it takes the elevator 20 seconds to travel from ground level to the top floor and that the total elevation change is 50 ft. With those assumptions, the average power is

$$P_{avg} = \frac{(500 \text{ lb} + 2500 \text{ lb})(50 \text{ ft})}{20 \text{ s}} = 7500 \text{ ft·lb/s}$$

Converting to the unit of horsepower,

$$P = (7500 \text{ ft·lb/s})(1.818 \times 10^{-3} \text{ hp/(ft·lb/s)}) = 13.6 \text{ hp}$$

As an engineer later accounts for friction, inefficiency of the motor, the possibility of the elevator being overloaded, and a factor of safety, the calculation would become more accurate. From a preliminary standpoint, however, a motor rated for a few tens of horsepower, rather than one with several horsepower or several hundred, would be sufficient.   ∎

6.3 HEAT AS ENERGY IN TRANSIT

In the previous section, we considered several types of energy that can be stored within a mechanical system. Energy can also be converted from one form to another—for instance, when the potential energy of a falling object is transferred to its kinetic energy. In a similar manner, the chemical energy that is stored in a fuel is released as heat when it burns. We view heat as energy that is in transit from one location to another because of a temperature difference. In this section, we explore several engineering concepts that are related to heat, its release when a fuel burns, and its transfer through the processes known as conduction, convection, and radiation.

Heat of Combustion

When a fuel such as oil (liquid), coal (solid), or propane (gas) is burned, the chemical reactions that occur release heat and by-products including water vapor, carbon monoxide, and particulates. The chemical energy that is liberated as heat during combustion is measured by a quantity called the *heat of combustion*, *H*. As listed in Table 6.3, the numerical values for the heat of combustion describe the ability of fuels to produce heat, and *H* is specified in the units MJ/kg in the SI and Btu/slug in the USCS. Conversion factors between conventional units in the USCS and SI for heat of combustion are listed in Table 6.4. The unit Btu shown in Tables 6.1 and 6.3 is an abbreviation for the "British thermal unit." Historically, one Btu was defined as the quantity of heat that must be supplied in order to raise the

British thermal unit
- - - - - - - - - - - - - - ➔

TABLE 6.3 Heat of Combustion Values for Different Fuels

Type	Substance	Heat of Combustion, H MJ/kg	Btu/slug
Gas	Natural gas	25	3.5×10^5
Liquid	Methanol	23	3.2×10^5
	Ethanol	30	4.2×10^5
	Gasoline	45	6.2×10^5
	Fuel oil	42	5.8×10^5
Solid	Coal	28	3.9×10^5
	Wood	20	2.8×10^5

TABLE 6.4 Conversion Factors Between Units for Heat of Combustion in the USCS and SI

MJ/kg	Btu/slug
1	1.383×10^4
7.229×10^{-5}	1

temperature of 1 lb of water by 1° F. In the modern definition, the Btu is equivalent to 778.2 ft·lb or 1055 J. The Btu is a commonly used unit for heat and energy calculations in mechanical engineering.

As a general rule, the heat ΔQ that is released when a mass m of fuel is burned is given by

$$\Delta Q = mH \tag{6.6}$$

Referring to Table 6.3, when 1 kilogram of gasoline is burned, some 45 MJ of heat are released. Equivalently in the USCS, we say that 6.2×10^5 Btu are released when 1 slug of gasoline is consumed. If we could somehow build an automobile engine that perfectly converts that quantity of heat into kinetic energy, a 1000-kilogram vehicle could be accelerated to a speed of nearly 300 m/s, roughly the speed of sound at sea level.

EXAMPLE 6.5

A gasoline-powered internal combustion engine generates an average power output of 50 kW (about 67 hp in the USCS). Neglecting the inefficiency of the engine, calculate the volume of fuel that is consumed in 1 hour. Express your result in both the SI and USCS.

SOLUTION

By discounting inefficiency, we recognize that our calculation will underestimate the actual rate of fuel consumption. In 1 hour, the engine produces an energy output of

$$50 \text{ kW} = (50 \text{ kJ/s})(3600 \text{ s}) = 1.8 \times 10^5 \text{ kJ} = 180 \text{ MJ}$$

Referring to Table 6.3, the heat of combustion for gasoline is 45 MJ/kg. Applying Equation 6.6, the mass of fuel burned is

$$m = \frac{180 \text{ MJ}}{45 \text{ MJ/kg}} = 4 \text{ kg}$$

In Table 4.3, the density of gasoline is listed as 680 kg/m^3, and so the engine burns

$$V = \frac{4 \text{ kg}}{680 \text{ kg/m}^3} = 0.0059 \text{ m}^3 = 5.9 \text{ L}$$

where the conversion 1 m^3 = 1000 L from Table 2.4 was used. In the USCS and with the conversion factor 1 L = 0.2642 gal from Table 2.7, 1.6 gal of fuel is consumed in 1 hour. ∎

EXAMPLE 6.6

The average single-family household in the United States consumes 98 million Btu of energy each year. In the units of lb, what quantity of fuel oil must be burned to produce that amount of energy?

SOLUTION

The mass of fuel oil is found by applying Equation 6.6:

$$m = \frac{98 \times 10^6 \text{ Btu}}{5.8 \times 10^5 \text{ Btu/slug}}$$

from which $m = 170$ slugs. The fuel oil weighs

$$W = \text{mg} = (170 \text{ slugs})(32.2 \text{ ft/s}^2) = 5440 \text{ lb}$$ ∎

EXAMPLE 6.7

Referring to Example 6.3, the attack on the World Trade Center involved both the kinetic energy of the aircraft and the heat released as the jet fuel burned. With the approximation that the aircraft was loaded to its maximum fuel capacity of 90,000 L, calculate the amount of heat that was released. Use the following values for jet fuel: density 840 kg/m^3, and heating value 43 MJ/kg.

SOLUTION

Since the volume quantity of fuel is specified, we first determine its mass:

$$m = (840 \text{ kg/m}^3)(9 \times 10^4 \text{ L})(0.001 \text{ m}^3/\text{L}) = 7.56 \times 10^4 \text{ kg}$$

Equation 6.6 provides the released heat as

$$\Delta Q = (7.56 \times 10^4 \text{ kg})(43 \text{ MJ/kg}) = 3.25 \times 10^6 \text{ MJ} = 3250 \text{ GJ}$$

This is an enormous quantity, about a 1000 times greater than the kinetic energy of the plane itself as found in Example 6.3. The fuel released heat equivalent to some three-quarters of a kiloton, or over 1 million pounds, of high explosives. ∎

Specific Heat

Quenching
- - - - - - - - - - - - - - →

In its commercial production, hot steel is often rapidly cooled by immersing it in a bath of oil or water. The purpose of this processing step, known as *quenching*, is to harden the steel by modifying its internal structure. The material's ductility (Section 5.3) can be improved subsequently by a reheating operation called *tempering*. When a steel rod is held at, say, 800°C and then quenched in oil, heat flows from the steel and the oil bath becomes warmer. The energy that was stored within the steel decreases, and we observe that loss as a change in its temperature. Likewise, the temperature of the oil bath, and the energy stored within it, are raised.

Although the flow of heat between the steel and oil is intangible in the sense that we cannot directly see it take place, we can measure it by the temperature changes that occur. Heat is not the same as temperature, but changes in temperature do indicate that heat has been transferred. When the sun heats a driveway during the afternoon to 120°F or so, it remains warm well into the night. The large and massive driveway is able to store more energy than a pot of water that was warmed on a stove to the same temperature.

The ability of a system to accept heat and store it as internal energy depends on the type of material, the amount of it, and its change in temperature. Returning to the example of steel processing, consider a solid casting having mass m that is initially at temperature T_0. As heat ΔQ flows into the block, its temperature rises to the value T following the equation

$$\Delta Q = mc(T - T_0) \tag{6.7}$$

Known as specific heat, the parameter c is a property that describes how materials differ with respect to the amount of heat they must absorb in order to raise their temperature. Specific heat has the units of kJ/(kg·°C) in the SI and Btu/(slug·°F) in the USCS, and Table 6.5 lists numerical values for selected materials. For instance, in order to raise the temperature of a 1-kg sample of steel by 1°C, 0.45 kJ = 450 J of heat must be added to it. For changes in temperature that are not too large, and as long as the material does not change phase, it is acceptable to treat c as being constant. Conversion factors between

TABLE 6.5 Specific Heat Values of Different Materials

Type	Substance	Specific Heat, c	
		kJ/(kg·°C)	Btu/(slug·°F)
Solid	Aluminum	0.90	6.9
	Copper	0.39	3.0
	Steel	0.45	3.5
	Glass	0.84	6.5
Liquid	Ethanol	2.5	19
	Oil	1.9	15
	Water	4.2	32

TABLE 6.6 Conversion Factors Between Units for Specific Heat in the USCS and SI

kJ/(kg·°C)	Btu/(slug·°F)
1	7.685
0.1301	1

conventional units in the USCS and SI for specific heat are listed in Table 6.6, and you can convert temperature values between the Celsius and Fahrenheit scales using the equations

$$°F = (9/5)°C + 32 \qquad \text{and} \qquad °C = (5/9)(°F - 32) \tag{6.8}$$

EXAMPLE 6.8

A steel drill rod 8 mm in diameter and 15 cm long is being heat-treated in an oil bath. The 850°C rod is quenched and held at 600°C, and then quenched a second time to 25°C. Calculate the quantities of heat that must be removed at the two quenching stages.

SOLUTION

Table 5.2 lists the weight density of steel as 76 kN/m^3, which is equivalent to the mass density of 7750 kg/m^3. We first calculate the rod's mass:

$$m = \pi((0.008 \text{ m})^2/4)(0.15 \text{ m})(7750 \text{ kg/m}^3) = 0.058 \text{ kg}$$

By using Equation 6.7, the quantities of heat removed from the rod during the two quench stages are

$$\Delta Q = (0.058 \text{ kg})(0.45 \text{ kJ/(kg} \cdot {}^\circ\text{C}))(850{}^\circ\text{C} - 600{}^\circ\text{C}) = 6.53 \text{ kJ}$$

and

$$\Delta Q = (0.058 \text{ kg})(0.45 \text{ kJ/(kg} \cdot {}^\circ\text{C}))(600{}^\circ\text{C} - 20{}^\circ\text{C}) = 15.1 \text{ kJ} \qquad \blacksquare$$

EXAMPLE 6.9

Continuing with Examples 6.3 and 6.7, the structural steel of the World Trade Center was weakened by the high-temperature fire started by the burning jet fuel. At a temperature of 425°C, structural steel begins to weaken appreciably, and when heated near 650°C, it can support only about one-half of its original capacity. Approximating each floor in a tower of the World Trade Center as comprising 2 million kilograms of steel, calculate the number of floors of the building that could be raised to the 650°C point by the heat released from the fuel.

SOLUTION

In Example 6.7, combustion of the fuel was found to release 3250 GJ. Referring to Table 6.5, the specific heat of steel is 0.45 kJ/(kg·°C). Denoting the number of floors engulfed in fire by n and using the initial temperature of $T_0 = 25°C$, Equation 6.7 provides

$$3.25 \times 10^9 \text{ kJ} = n(2 \times 10^6 \text{ kg})(0.45 \text{ kJ/(kg} \cdot {}^\circ\text{C}))(650{}^\circ\text{C} - 25{}^\circ\text{C})$$

The calculation provides $n = 5.8$. On this basis, we conclude that sufficient heat was released from just the aircraft fuel to weaken the steel in at least several floors of a tower. Our estimate is conservative to the extent that the fires, once started, also consumed the paper, building materials, and other flammable items within the towers, only adding to the intensity of the fires and the heat released. $\qquad \blacksquare$

Transfer of Heat

We have described heat as energy that is being transferred from one location to another because of a temperature difference. The three mechanisms for heat transfer are known as *conduction*, *convection*, and *radiation*, and they arise in different mechanical engineering technologies.

Conduction

When you grip the hot handle of a pan on a stove, you feel conduction in action: Heat flows from the pan and along the length of the handle to its cooler free end. That process is illustrated by the metallic rod sketched in Figure 6.3. One end of the rod is held at the high temperature T_h, and the other end has the lower temperature T_l. Although the rod itself doesn't move, heat flows down it like

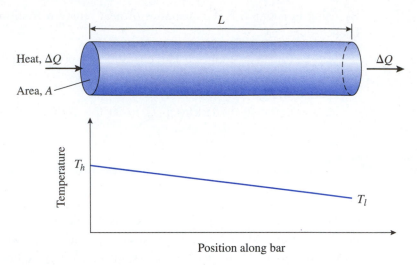

FIGURE 6.3 Heat conduction along a metallic bar.

a current because points along the rod have different temperatures. Just as a change in voltage produces an electrical current, the change in temperature, $T_h - T_l \neq 0$, must be present for heat conduction to take place.

The quantity of heat that flows along the rod in Figure 6.3 during the time interval Δt is

$$\Delta Q = \frac{\kappa A \Delta t}{L}(T_h - T_l) \tag{6.9}$$

This principle is known as *Fourier's law of heat conduction*, and it is named in honor of the French scientist and mathematician Jean Baptiste Joseph Fourier (1768–1830). Heat conduction occurs in proportion to the cross-sectional area A, and it is inversely proportional to the rod's length L. As introduced in Problem 15 in Chapter 2, the material property κ (the lowercase Greek character kappa) is called the *thermal conductivity*, and values for various materials are listed in Table 6.7.

Thermal conductivity
‑ ‑ ‑ ‑ ‑ ‑ ‑ ‑ ‑ ‑ ‑ ‑ ‑ ‑ ‑ ‑ ‑ →

When the thermal conductivity is large, heat flows through the material quickly. Metals generally have large values of κ, and thermal insulators such as fiberglass have low values. With reference to Table 6.7, 200 J of heat will flow each second through a square panel of aluminum that is 1 meter on a side, and 1 meter thick when the temperature difference across the panel's two faces is 1°C. Even among metals, the numerical values for κ can vary significantly. The thermal conductivity of aluminum is $4\frac{1}{2}$ times larger than the value for steel, and κ for copper is nearly 2 times larger still. Aluminum and copper are preferred metals for use in cookware for precisely that reason: Heat can more easily flow around a pan to prevent hot spots from forming and food from being burned. Conversion factors between conventional units in the USCS and SI for thermal conductivity are listed in Table 6.8.

TABLE 6.7   Thermal Conductivity of Different Materials

Material	Thermal conductivity, κ W/(m·°C)	(Btu/h)/(ft·°F)
Steel	45	26
Copper	390	220
Aluminum	200	120
Glass	0.85	0.50
Wood	0.3	0.17

TABLE 6.8   Conversion Factors Between Units for Thermal Conductivity in the USCS and SI

W/(m·°C)	(Btu/h)/(ft·°F)
1	0.5778
1.731	1

EXAMPLE 6.10

A small office has a 3-ft by 4-ft window on one wall. The window is made from single pane glass that is $\frac{1}{8}$-in. thick. While evaluating the building's heating and ventilation system, an engineer needs to calculate the heat loss through the window on a winter day. Thermocouples are used to measure the temperature of the glass on the inside and outside surfaces of the window, and the outside is found to be 3°F cooler. What quantity of heat is lost through the window each hour?

SOLUTION

The thermal conductivity of glass is listed in Table 6.7 as 0.50 (Btu/h)/(ft·° F). To keep the units in Equation 6.9 consistent, we first convert the thickness as follows:

$$L = (0.125 \text{ in.})(0.0833 \text{ ft/in.}) = 0.0104 \text{ ft}$$

The heat loss in one hour becomes

$$\Delta Q = \frac{[0.50 \text{ (Btu/h)/(ft·°F)}](3 \text{ ft})(4 \text{ ft})(1 \text{ h})}{0.0104 \text{ ft}}(3°F) = 1730 \text{ Btu}$$

Using Table 6.1, that rate of heat loss is equivalent in the SI to

$$(1730 \text{ Btu/h})(1055 \text{ J/Btu})(2.778 \times 10^{-4} \text{ h/s}) = 507 \text{ J/s} = 507 \text{ W}$$

We conclude that a small electric heater in the 500-W range would be sufficient to compensate for that rate of heat loss. ∎

Convection
- - - - - - - - - - - - - ►

In addition to conduction, heat can also be transferred between locations by a fluid that is in motion; that process is known as *convection*. The cooling system of an automobile engine, for instance, operates by pumping a mixture of water and antifreeze through passageways inside the engine block. Excess heat is removed from the engine, transferred temporarily to the coolant by convection, and ultimately released into the air by the vehicle's radiator. Because a pump

Forced convection
- - - - - - - - - - - - - ►

causes the coolant to circulate, heat transfer is said to occur by forced convection. In other circumstances, a liquid or gas can circulate on its own, without a pump or fan being present, because of the buoyancy forces created by temperature variations within the fluid. As air is heated, it becomes less dense, and buoyancy forces cause it to rise and circulate. The rising flow of warm fluid,

Natural convection
- - - - - - - - - - - - - ►

and the falling flow of cooler fluid to fill its place, is called *natural convection*. So-called thermals develop near mountain ridges and bodies of water, and they are natural convection currents in the atmosphere that sailplanes, hang gliders, and birds exploit to stay aloft. In fact, many aspects of the Earth's climate, oceans, and molten core are related to natural convection. Giant convection cells (some the size of Jupiter) are even present in the gases of the Sun, and they interact with its magnetic field to influence the formation of sunspots.

Radiation
- - - - - - - - - - - - - ►

The third mechanism of heat transfer is called *radiation*. The term refers to the characteristic that heat is emitted and absorbed without direct physical contact, and it is unrelated to "radiation" in the context of nuclear processes or power generation. Radiation occurs when heat is transmitted by the long infrared waves of the electromagnetic spectrum. Those waves are able to propagate through air, and even through the vacuum of space. Energy from the Sun, for instance, reaches the Earth through the process of radiation, and as those electromagnetic waves are absorbed by air, land, and water, they are converted to heat. The radiators in a home heating system comprise winding metal pipes through which steam or hot water circulates. If you place your hand directly on the radiator, you feel heat flow directly into your hand by conduction. However, even if you stand some distance away and do not touch the radiator, you are still warmed by it through the process of radiation.

6.4 ENERGY CONSERVATION AND CONVERSION

Now that we have discussed the basic concepts of energy, work, and heat, in the next two sections we explore the physical laws governing energy conservation and conversion. Engineering analyses are built upon the concept of a system, defined as a collection of materials and components that are examined together with respect to their thermal and energy behavior. The system is imagined to be enclosed entirely within a surface that separates it from the surroundings and any extraneous effects. This viewpoint is analogous to our approach in Chapter 4 when constructing free body diagrams—all forces that crossed an imaginary boundary were included on the diagram. In a similar manner, engineers analyze thermal and energy systems by considering (1) heat that flows in or out of the system, (2) work that the system performs on the surroundings or vice versa, and (3) energy changes within the system.

Consider the sketch of a generic thermal and energy system in Figure 6.4. The quantity of heat ΔQ flows into the system, and work ΔW is performed by it as an output. The internal or potential energy of the system might also change by the amount ΔU. The first law of thermodynamics states that these three quantities balance according to

First law

$$\Delta Q = \Delta W + \Delta U \qquad (6.10)$$

In applying this equation, it is necessary to have a sign convention and to be consistent when applying positive and negative signs to each quantity. For instance, ΔQ is positive when heat flows into the system; ΔW is positive when the system performs work on its surroundings; and ΔU is positive when the system's internal energy increases.

In addition to the requirement of energy conservation, Equation 6.10 also implies that an equivalence exists between heat, work, and energy and that one form can, in principle, be exchanged for another. For instance, the first law describes how an internal combustion engine operates, where the heat ΔQ that is released by burning gasoline is converted into mechanical work ΔW and the vehicle's kinetic energy. The first law establishes an opportunity for many other devices in mechanical engineering, ranging from air conditioners, to jet engines, to the power plants used for generating electricity. Left aside until Section 6.5 are the limitations that restrict how efficiently heat, work, and energy can be exchanged. The following example illustrates a calculation involving energy conservation and the conversion of kinetic energy into thermal energy.

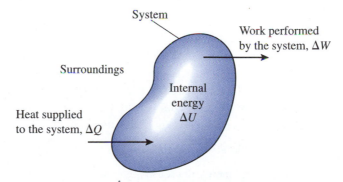

FIGURE 6.4   Schematic of the first law energy balance for a generic thermal and energy system.

EXAMPLE 6.11

The driver of a 1200-kg automobile traveling at 100 km/h applies the brakes and the vehicle comes to a complete stop. The automobile's initial kinetic energy is converted by the braking system into heat. In turn, the temperatures of the cast-iron brake rotors and drums ($c = 0.43$ kJ/(kg·°C)) increase. The vehicle has front disk and rear drum brakes, and the system is balanced so that the front set provides 75% of the total braking capacity (Figure 6.5). What is the temperature rise of the two 7-kg brake rotors?

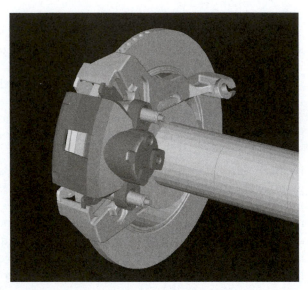

FIGURE 6.5 Computer analysis of the rotor, caliper, and pads in an automobile front disk brake system, as described in Example 6.11.
Source: Reprinted with permission of Mechanical Dynamics, Inc.

SOLUTION

As the vehicle comes to a stop, a portion of its initial kinetic energy is lost through air drag, rolling resistance of the tires, and wear of the brake pads and shoes, but we neglect those factors at the first level of approximation. We balance the decrease of the automobile's kinetic energy with the energy stored within the brake rotors as their temperature rises. In the units of m/s, the velocity is

$$v = (100 \text{ km/h})(1000 \text{ m/km})(2.778 \times 10^{-4} \text{ h/s}) = 27.8 \text{ m/s}$$

and the kinetic energy decreases by

$$\Delta U = \frac{1}{2}(1200 \text{ kg})(27.8 \text{ m/s})^2 = 4.63 \times 10^5 \text{ N·m} = 463 \text{ kJ}$$

Because the front brakes provide three-quarters of the braking capacity, $(0.75)(463 \text{ kJ}) = 347 \text{ kJ}$ of heat flows into the rotors. Their temperature rise is found by applying Equation 6.7:

$$347 \text{ kJ} = 2(7 \text{ kg})(0.43 \text{ kJ/(kg·°C)})\Delta T$$

where the factor of 2 accounts for both rotors. The calculation indicates that the temperature of the rotors increases by $\Delta T = 58°C$. ■

In the following example, we consider the conversion of potential energy into electrical energy in the context of a hydroelectric power plant. The United States is the world's leading producer of hydroelectric power, a source that supplies some 13% of the nation's total electricity needs. Figures 6.6 and 6.7 depict the Hoover Dam, which is capable of generating over 2 GW of power. The dam's turbines are fed by water stored in a lake containing approximately enough water to cover the Commonwealth of Pennsylvania to a depth of 1 foot. In a hydroelectric power plant, water stored in such a large reservoir falls downward through passageways or conduits called penstocks. Located near the dam's base, each turbine is essentially a high-performance water

■ **FIGURE 6.6**   The Hoover Dam viewed from a position downstream of its face.

Source: Reprinted with permission of the Bureau of Reclamation, United States Department of the Interior.

FIGURE 6.7 Generators in the Hoover Dam convert rotation of its turbine shafts into electrical power. The stairs and railings give scale to these massive machines.
Source: Reprinted with permission of the Bureau of Reclamation, United States Department of the Interior.

wheel, and they are driven to rotate by the high-speed flow of water. The turbines are connected by shafts to generators that produce the electrical power. Once the water exits the turbines, its speed has been significantly reduced, and it is discharged into a river.

EXAMPLE 6.12

In the diagram of a hydroelectric power plant in Figure 6.8, the vertical drop of water from the reservoir to the turbines is 100 m, and water flows through the system and into the lower river at the rate of 500 m³/s. Discounting friction and inefficiency of the turbines and generators, how much electrical energy can be produced each second? Neglect the water's relatively small velocity as it enters the top of the penstocks and as it exits the turbines. Use the value 1000 kg/m³ for the density of water.

SOLUTION

The potential energy of water in the reservoir is converted into kinetic energy as it falls, and in turn, the water's kinetic energy is transferred to rotation of the turbines and

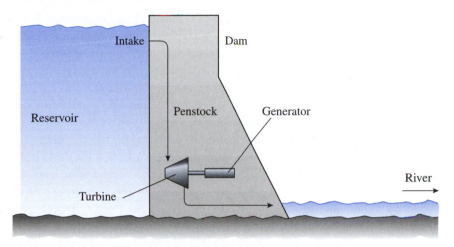

FIGURE 6.8   Schematic of the dam and power plant in Example 6.12.

generators. We neglect the components of the water's kinetic energy that correspond to its velocity when flowing out of the reservoir into the penstocks and from the turbines into the river. Each is small when compared to the overall change in gravitational potential energy. The mass of water that exits the reservoir, flows through the penstocks and turbines, and discharges into the lower river each second is

$$m = (500 \text{ m}^3)(1000 \text{ kg/m}^3) = 5.0 \times 10^5 \text{ kg}$$

The corresponding decrease in the reservoir's store of gravitational potential energy is

$$\Delta U = (5.0 \times 10^5 \text{ kg})(9.81 \text{ m/s}^2)(100 \text{ m}) = 4.9 \times 10^8 \text{ N·m} = 490 \text{ MJ}$$

With ideal efficiency, 490 MJ of energy can be transferred to the generators and converted into electrical energy each second. By using the definition of average power in Equation 6.5 with $\Delta t = 1$ s, the plant's power output could be as high as 490 MW. Owing to friction in the piping system and inefficiency of the turbines and generators, however, an actual hydroplant would necessarily have a lower power output. ■

Just as we discussed fluid buoyancy, drag, and lift forces in Chapter 4, we next consider the special form that the principle of energy conservation takes when it is applied to a liquid or gas. This principle is named after the eighteenth-century mathematician and physicist Daniel Bernoulli. It can be applied when no energy is dissipated because of friction or the fluid's viscosity; no work is performed either on the fluid *Bernoulli's equation* (for instance, by a pump) or by it (in a turbine); and no heat transfer takes place. Together, those restrictions frame the flowing fluid as a conservative energy system, and Bernoulli's equation becomes

$$\frac{p}{\rho} + \frac{v^2}{2} + gh = \text{constant} \tag{6.11}$$

Here p and ρ are the pressure and density of the liquid or gas as described in Section 4.5, v is its speed, g is the gravitational acceleration constant, and h is the height of the fluid above a reference point. Numerical values for ρ in both the USCS and SI are listed in Table 4.3 for various liquids and gases. The three terms on the left-hand side of the equality in Equation 6.11 represent the work of pressure forces, the kinetic energy of the steadily flowing fluid, and its gravitational potential energy. Bernoulli's equation is dimensionally consistent, and each quantity in it has the units of energy-per-unit-mass.

Venturi flow meter
- - - - - - - - - - - - - - ▶

One implication of Equation 6.11 is that a trade-off exists between the pressure, velocity, and elevation of a flowing fluid. In a device called a venturi flow meter, mechanical engineers are able to exploit that equivalence and determine the velocity of a liquid or gas by measuring only its pressure (or more precisely, the change in pressure between two points). To understand how a venturi flow meter works, Figure 6.9(a) illustrates water passing through a pipe. At the constriction, the cross-sectional area of the pipe reduces from A_1 to the smaller value A_2. The curves of Figure 6.9(a) are called *streamlines*, and they help us to visualize the water's velocity and flow pattern at each point in the pipe. The pressure sensors shown in Figure 6.9(b) are located just before and after the constriction, and the change $p_1 - p_2$ that they measure is enough information to determine the water's velocity.

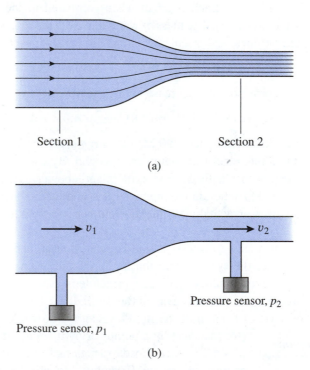

Section 1   Section 2

(a)

v_1 v_2

Pressure sensor, p_2

Pressure sensor, p_1

(b)

■ **FIGURE 6.9**   (a) A liquid is pumped through a pipe with a constriction. The cross-sectional areas of the pipe at the two sections are A_1 and A_2. (b) A venturi flow meter is used to measure the fluid's velocity.

The central streamline in Figure 6.9(a) does not change its vertical elevation, and so the potential energy term in Equation 6.11 remains the same before and after the constriction. The pressure and velocity of the water at the two cross sections balance according to

$$\frac{p_1}{\rho} + \frac{v_1^2}{2} = \frac{p_2}{\rho} + \frac{v_2^2}{2} \tag{6.12}$$

which we rearrange to read

$$v_2^2 - v_1^2 = \frac{2}{\rho}(p_1 - p_2) \tag{6.13}$$

The volume of water that flows through the pipe per unit time is the product of the cross-sectional area of the pipe and the velocity of the water. For instance, if the pipe area is $A = 1.5 \text{ ft}^2$ and water flows through it at $v = 3$ ft/s, the volume flow rate is $Av = 4.5 \text{ ft}^3$/s or 33.7 gal/s, where the conversion 1 gal $= 0.1337 \text{ ft}^3$ from Table 2.2 has been applied. Because water doesn't build up or accumulate in the pipe, the amount of water that flows into the first section must also exit the second section, and the two flow rates are identical:

$$A_1 v_1 = A_2 v_2 \qquad \text{or} \qquad v_1 = (A_2/A_1)v_2 \tag{6.14}$$

Combining Equations 6.13 and 6.14, the velocity of the water after the constriction can be calculated directly from its pressure drop:

$$v_2 = \sqrt{\frac{2(p_1 - p_2)}{\rho(1 - (A_2/A_1)^2)}} \tag{6.15}$$

As water flows through the bottleneck in Figure 6.9, its velocity increases and the pressure drops. The "venturi effect" is the principle behind the operation of such hardware as the valves and nozzles in automobile carburetors and fuel injection systems and the aspirators that deliver pharmaceuticals by inhalation to medical patients.

The lift force developed by an airfoil, which we introduced in Chapter 4, can also be understood in the context of Equation 6.11. For flow around the airfoil in Figure 6.10(a), the gravitational potential energy term gh in Equation 6.11 can be neglected because the elevation changes are small compared to the other terms present. The quantity $(p/\rho) + (v^2/2)$ is therefore approximately constant as air moves along the airfoil's upper and lower surfaces. As the air speeds up and flows above the airfoil, its pressure decreases by a corresponding amount in accordance with Equation 6.11. The airfoil's lift force is produced by the pressure difference between its (lower-pressure) top surface and (higher-pressure) bottom surface. The results from a computer simulation in Figure 6.10(b) show in more detail how the air pressure is distributed around a wing.

Pitot tube

Another application of Equation 6.11 to flight is a device called the *pitot tube*, which functions as an aircraft's speedometer by measuring air velocity. The pitot tube is often mounted on a boom that extends off the aircraft's nose or wing, and it points forward in the direction of flight. As shown in Figure 6.11, the pitot tube is constructed as a hollow pipe with two internal passageways. The first

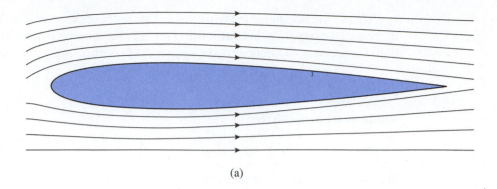

(a)

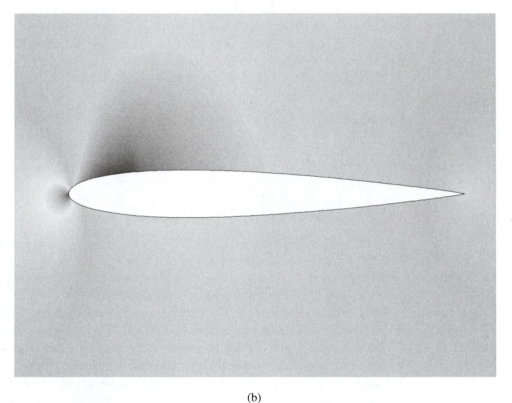

(b)

■ **FIGURE 6.10**   (a) Airflow around the cross section of a wing. (b) Results from a computer simulation of an airfoil depicting the low-pressure region on the upper surface. *Source:* Reprinted with permission of Fluent Inc.

pressure sensor connects to a hole that runs down the center of the pipe, and the second pressure tap is taken from the outside surface. By measuring the air pressure difference between the point at the tip of the tube and a point on its side, an accurate airspeed reading can be obtained. Because the tip of the pitot tube is a point of symmetry where

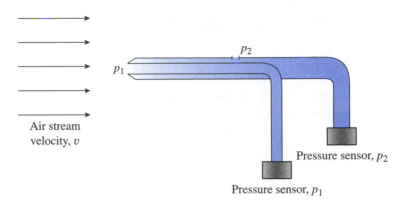

FIGURE 6.11   A pitot tube is used to measure the velocity of an airstream.

air flows equally upward and downward, the velocity of the airstream is zero at that location. Application of Equation 6.11 provides the velocity of the airstream as

$$v = \sqrt{\frac{2(p_1 - p_2)}{\rho}} \tag{6.16}$$

where the pressure difference is measured by the electronic sensors and ρ is the density of air in Table 4.3. If v is not well below the speed of sound, Equation 6.16 should not be applied because certain assumptions behind the analysis would be violated. In particular, air compresses significantly as the Mach number grows, and at supersonic speeds, shock waves (such as those shown in of Figure 4.22) will form off the tip of the pitot tube.

EXAMPLE 6.13

A commuter aircraft travels at 260 mph. In the units of psi, what pressure difference is measured by the pitot tubes mounted on its wings?

SOLUTION

Referring to Table 4.3, the density of air is 2.33×10^{-3} slug/ft³. In consistent units, the air speed is

$$v = (260 \text{ mph})(5280 \text{ ft/mi})(2.778 \times 10^{-4} \text{ h/s}) = 382 \text{ ft/s}$$

By applying Equation 6.16, the pressure difference becomes

$$\Delta p = p_1 - p_2 = \frac{1}{2}(2.33 \times 10^{-3} \text{ slug/ft}^3)(382 \text{ ft/s})^2 = 170 \text{ lb/ft}^2$$

With the conversion factor 1 psf $= 6.944 \times 10^{-3}$ psi, we calculate the pressure difference to be 1.18 psi. Although that differential is only about a twelfth of an atmosphere, it can be easily measured by electronic pressure gauges.   ∎

6.5   HEAT ENGINES AND EFFICIENCY

An important aspect of engineering is designing machinery that produces mechanical work from a fuel. At the simplest level, natural gas or fuel oil can be burned, and the heat that is released can be used to warm a house or other building. Of equal practical importance is the need to take the next step and produce work from that heat. Mechanical engineers are concerned with the efficiency of systems in which a fuel is burned, heat is released, and a machine converts that heat into useful work. By increasing the efficiency of that process, it is possible to improve the fuel economy of an automobile, increase its power output, and reduce the engine size. In this section, we discuss such energy conversion and power generation systems, and we develop the concepts of real and ideal efficiency.

Heat engine

The heat engine sketched in Figure 6.12 is any device that converts the heat supplied to it into mechanical work. As the input, the engine absorbs a quantity of heat, Q_h, from the high-temperature heat reservoir, which is maintained at the constant temperature T_h. A portion of Q_h is converted into mechanical work W by the engine. The remainder of the heat is exhausted from the engine as a waste

Heat reservoir

product. The lost heat Q_l is released into the low-temperature reservoir maintained at the constant value $T_l < T_h$. From this conceptual viewpoint, the heat reservoirs are large enough so that their temperatures do not change as heat is removed or added.

To fix ideas in the context of an automobile engine, Q_h represents the heat released by burning fuel in the engine's combustion chamber; W is the mechanical work associated with rotation and torque of the crankshaft; and Q_l is the waste heat

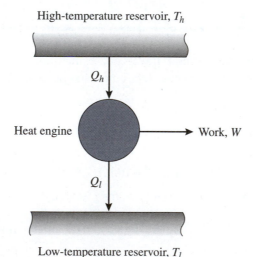

FIGURE 6.12   Conceptual view of a heat engine that operates between high- and low-temperature heat reservoirs.

that is lost from the exhaust pipe, that warms the engine block, and that warms the passenger compartment. The engine is not able to completely convert the supplied heat into mechanical work, and some fraction is wasted and released into the environment. Heat is transferred to the engine at the gasoline combustion temperature, labeled as T_h in Figure 6.12, and the wasted heat Q_l is released at the lower temperature T_l. Alternatively, as applied to an electrical power plant, Q_h would represent the heat produced by burning a fuel such as coal, oil, or natural gas. The energy released by it powers the plant's turbines and generators to produce electricity, but the plant also releases some waste heat Q_l into the atmosphere from cooling towers or into a nearby river, lake, or ocean. With W being regarded as the useful output, all thermal and energy systems such as these have the characteristic that the supplied heat cannot be entirely converted into work.

Referring to Equation 6.10 and Figure 6.12, the energy balance of the engine is

$$Q_h - Q_l = W \tag{6.17}$$

Efficiency - - - - - - - - ▶

because there is no internal energy change. The real efficiency η (the lower-case Greek character eta) of the heat engine is defined as the ratio of work to the heat that was supplied:

$$\eta = \frac{W}{Q_h} = \frac{Q_h - Q_l}{Q_h} = 1 - \frac{Q_l}{Q_h} \tag{6.18}$$

Because $Q_h > Q_l$, the efficiency always lies between 0 and 1. Efficiency is sometimes described as the ratio of "what you get" (the work) to "what you paid for" (the heat input), and it is often expressed as a percent. If an automobile engine has an overall efficiency of 25%, then for every 4 gallons of fuel consumed, 1 gallon's worth of energy would be converted for the useful purpose of powering the automobile. Such would be the case even for a heat engine that has been optimized by millions of hours of engineering research and development.

Equation 6.10 sets an upper limit on the amount of work that can be obtained from a heat source: It is not possible to obtain more work from a machine than the amount of heat that was supplied to it. Even that view is unrealistically optimistic in the sense that no one has yet been able to develop a machine that converts heat into work perfectly. Efficiency levels of 25% for an automobile engine or 30% for an electrical power plant are realistic, and those values reflect practical limits on everyday

Second law - - - - - - - - ▶

heat engines. The second law of thermodynamics formalizes that observation and expresses the impossibility of developing an engine that does not waste heat:

> No device can operate in a cycle and only transform the heat supplied to it into work without also rejecting some portion of the supplied heat.

If the converse of this statement were valid, you could design an engine that would absorb heat from the air and power an aircraft or automobile without consuming any fuel. Likewise, you could build a submarine powered only by heat extracted from the ocean. In effect, you would be able to get something (the kinetic energy of the automobile, aircraft, or submarine) for nothing (because no fuel would be needed). Our practical experience, of course, is that such devices are not possible and that efficiencies are never ideal. While the first law stipulates that energy is conserved and can be converted, the second law places restrictions on the manner in which energy can be used.

Given that ideal conversion of heat into work is not possible, what is the maximum efficiency that a heat engine can reach? That important upper bound was established by the French engineer Sadi Carnot (1796–1832), who was interested in building engines that produced the greatest work from a given amount of steam. The *Carnot cycle* refers to a heat engine that has the highest possible efficiency within the constraints set by physical laws. Although an engine that operates precisely on this cycle cannot be built, it nevertheless provides a useful point of comparison when evaluating and designing real engines. The cycle is based on expanding and compressing an ideal gas in a cylinder and piston mechanism. At various stages of the cycle, heat is supplied to the gas, work is performed by the gas, and heat is rejected. The gas, cylinder, and piston form a heat engine that operates between reservoirs at the high and low temperatures of T_h and T_l, and it is ideal in the sense that the efficiency

Carnot cycle

Carnot efficiency

$$\eta_c = 1 - \frac{T_l}{T_h} \tag{6.19}$$

Rankine

Kelvin

is attained. In calculating the ratio of temperatures, T_h and T_l must be expressed on the same absolute scale, either in degrees Kelvin (°K) in the SI or degrees Rankine (°R) in the USCS. Absolute temperatures are converted from values on the familiar Celsius and Fahrenheit scales through

$$°K = °C + 273 \qquad °R = °F + 460 \tag{6.20}$$

The efficiency calculated from Equation 6.19 is called the *ideal* or *Carnot value*, and it does not depend on the internal details of the engine's construction. Performance is determined entirely by the temperatures of the heat reservoirs. The efficiency can be increased by lowering the temperature at which heat is exhausted or by raising the temperature at which heat is supplied. However, the practical limitations associated with the temperature at which fuel burns, the melting point of the metals making up the engine components, and the temperature of our environment restrict the latitude that engineers have in raising T_h or lowering T_l. Further, the realistic inefficiencies associated with friction, fluid viscosity, and other losses are not considered by η_c. You should view the Carnot efficiency as a limit based on energy conversion principles, and as such, it sets the upper bound on the real efficiency η of Equation 6.18.

In the following example, we apply Equation 6.19 to analyze the feasibility of a system for generating electricity from solar power. On a clear day with the sun directly overhead, some 1 kW of sunlight strikes each square meter of ground, and mechanical engineers recognize the potential for using that power. Approximately 250,000 homes in the United States already produce some portion of their electricity from photovoltaic cells. Used in applications ranging from handheld calculators to communication space-craft, solar cells are one technique for producing electricity from sunlight. However, solar cells are a relatively expensive way to generate electricity because they are made from processed silicon, just like integrated circuits and computer microprocessors. Because of their expense and efficiencies of about 15% when mass-produced, photo-voltaic cells are not presently well suited for applications involving large quantities of electrical power. In a heat engine known as a dynamic solar power generator, a parabolic mirror is arranged to capture, reflect, and focus sunlight onto a pipe that runs along the mirror's focus. Water flows through the pipe, is boiled into high-temperature steam, and drives a turbine and electrical generator. We next evaluate the efficiency of such a solar power generator, as sketched in Figure 6.13.

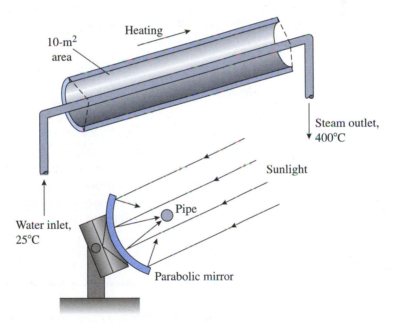

■ **FIGURE 6.13**   The dynamic solar power generation system uses a network of parabolic mirrors shaped like troughs to focus sunlight and convert water into high-temperature steam.

EXAMPLE 6.14

Solar energy is collected by parabolic mirrors and used to heat steam to the temperature $T_h = 400°C$. The steam powers turbines and generators, and the system releases its waste heat into the air at $T_l = 25°C$. Using the best-case solar radiation constant of $1\,\text{kW/m}^2$, estimate the number of 10-m^2 mirrors that will be needed to produce an average of 10 MW of power over a 24-hour period.

SOLUTION

For our calculation, we recognize that the plant can operate only during the peak daylight hours, roughly one-third of a day. Thus, to have an average output of 10 MW, the plant must be sized to produce 30 MW as it operates during daylight. Further, a method to store the excess power during the day and to retrieve it at night would also need to be devised. The real efficiency of the power plant cannot be greater than the Carnot efficiency

$$\eta_c = 1 - \frac{(25 + 273)°\text{K}}{(400 + 273)°\text{K}} = 0.56 = 56\%$$

Here we have converted temperatures to the absolute Kelvin scale as required in Equation 6.19. If the output power is 30 MW, heat must be supplied at the rate of $(30\,\text{MW})/(0.56) = 53.6\,\text{MW}$. In terms of the solar radiation constant, the plant will need to collect sunlight over the area

$$A = \frac{53.6 \times 10^3\ \text{kW}}{1\ \text{kW/m}^2} = 5.36 \times 10^4\ \text{m}^2$$

which is equivalent to a square parcel of land 230 m on a side. This system would require some 5360 individual 10-m^2 collector mirrors. In our calculation, we have neglected the factors of cloud cover, atmospheric humidity, and geographical latitude, each of which would decrease the amount of solar radiation that strikes the mirrors. The power plant footprint would need to be larger than our estimate suggests because the real efficiency will be lower than the Carnot value, but from a preliminary design standpoint, the calculation begins to firm up the plant's size and specifications. ∎

EXAMPLE 6.15

The technology of ocean thermal energy conversion uses the temperature difference between the (warm) water on the surface of an ocean and the (cold) deep water, to drive an electrical power generation plant. As long as a temperature difference of approximately 40°F exists between the surface and the deep water, net power can be generated once the overhead for pumps and other equipment has been subtracted. Consider a small demonstration plant in the equatorial Pacific Ocean that uses surface water at 90°F and cool 40°F deep water. How many gallons of seawater must be processed through the system each second in order to produce a gross output of 200 kW? The weight density of seawater is 8.6 lb/gal, and its specific heat is 95% that of freshwater.

SOLUTION

We recognize that the power plant will be less efficient than a Carnot cycle. When applying Equation 6.19, our objective is to obtain an upper-bound estimate of the efficiency. The heat engine operates between the absolute temperatures of $T_h = (90 + 460)°R = 550°R$ and $T_l = (40 + 460)°R = 500°R$, expressed on the Rankine scale. The ideal efficiency is

$$\eta_c = 1 - \frac{500°R}{550°R} = 0.09 = 9\%$$

With that low value and the target of 200 kW, the plant must receive heat at the rate of (200 kW)/(0.09) = 2220 kW. During 1 s, the input is (2220 kJ)(0.9478 Btu/kJ) = 2104 Btu, which is obtained from the energy stored in the water and associated with the 50°F temperature difference. Referring to Table 6.5 and the given information, the specific heat for seawater is 30.4 Btu/(slug·°F). By using Equation 6.7,

$$2104 \text{ Btu} = m(30.4 \text{ Btu/(slug·°F)})(50°F)$$

from which we see that the plant must process $m = 1.38$ slugs, or 44.6 lb, of seawater each second. Using the density of seawater for conversion, the ocean thermal energy system must process (44.5 lb)/(8.6 lb/gal) = 5.2 gal per second. The volume flow rate is modest in this case because of the high specific heat of the seawater. ∎

6.6 CASE STUDY #1: INTERNAL COMBUSTION ENGINES

In the remaining sections of this chapter, we turn our attention to three technologies of mechanical engineering that are based on the efficient conversion of energy from one form to another: internal combustion engines, electrical power generation, and aircraft jet engines.

The internal combustion engine is a heat engine that converts the chemical energy stored by gasoline (and fuels such as diesel, methanol, or propane) into mechanical work. The heat that is generated by rapidly burning a mixture of fuel and air in the engine's combustion chamber is transformed into rotation of the engine's crankshaft at a certain speed and torque. As you know, the applications of internal combustion engines are wide and varied, and they include automobiles, motorcycles, aircraft, ships, pumps, and electrical generators. Mechanical engineers develop extensive computer models of internal combustion engines, and before a single engine is even built, performance will be analyzed from the perspectives of fuel economy, power-to-weight ratio, noise, air pollution, and cost. In this section, we discuss the design, terminology, and energy principles behind four-stroke and two-stroke engines.

As shown in Figure 6.14, the main elements of a single-cylinder engine are the piston, cylinder, connecting rod, and crankshaft. Those components convert the back-and-forth motion of the piston into crankshaft rotation; in Chapter 7, we will examine

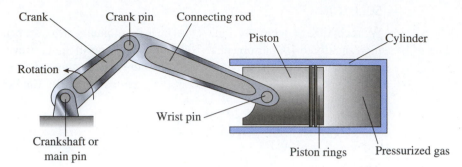

■ **FIGURE 6.14** Layout of the mechanism in a single-cylinder internal combustion engine. For clarity in the drawing, the components have not been drawn to scale.

the motion of that mechanism in detail. As fuel burns, the high pressure that develops in the cylinder pushes the piston, moves the connecting rod, and rotates the crankshaft. The engine also contains a means for fresh fuel and air to be drawn into the cylinder and for exhaust gas to be vented away. We will discuss those processes separately for the cases of four-stroke and two-stroke engines.

■ **FIGURE 6.15** This four cylinder, sixteen-valve automobile engine produces a peak power of 100 kW.

Source: © 2001 General Motors Corporation. Used with permission of GM Media Archives.

While the single cylinder configuration of Figure 6.14 may be the simplest, the power output is limited by its small size. In multicylinder engines, the pistons and cylinders are placed in the vee, in-line, or rotary orientations. For instance, four-cylinder engines with the cylinders arranged in a single straight line are common in automobiles (Figure 6.15). In a V-6 or V-8 engine, the engine block is made short and compact by setting the cylinders in two banks of three or four cylinders each. The angle between the banks is usually in the range 60°–90°, and in the limit of 180°, the cylinders are said to be horizontally opposed. Engines as large as V-12s and V-16s are found in heavy-duty trucks and luxury vehicles, and 54-cylinder engines comprising six banks of nine cylinders each have been used in some naval applications. The power produced by an internal combustion engine depends not only on the number of cylinders but also on its throttle setting and speed. Each automobile engine, for instance, has a particular speed where the output power is the greatest. An engine that is advertised as generating, say, 200 hp does not do so under all operating conditions. As an example, Figure 6.16 shows the power–speed performance curve of a V-6 automobile engine. This engine was tested at full throttle and reached its peak power output near a speed of 6000 revolutions per minute.

Power curve
- - - - - - - - - - - ➤

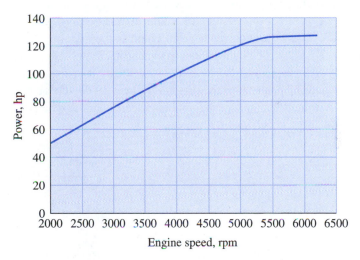

■ **FIGURE 6.16**   Measured power output of a 2.5-liter automobile engine as a function of its speed.

Four-Stroke Engine Cycle

Engine valve
- - - - - - - - - ➤

The cross section of a single-cylinder four-stroke engine is illustrated in Figure 6.17 at each stage of its operation. The engine is shown with two *valves* per cylinder: one for intake of fresh fuel and air, and one for exhaust of the combustion waste products. The mechanism that causes the valves to open and

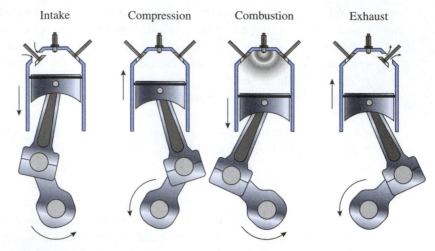

Intake   Compression   Combustion   Exhaust

FIGURE 6.17   Stages of the four-stroke engine cycle.

close is an important aspect of this type of engine, and Figure 6.18 depicts one design for an intake or exhaust valve. The valve closes as its head contacts the polished surface of the cylinder's intake or exhaust port. A specially shaped metal lobe, called a *cam*, rotates and controls the opening and closing motions of the valve. The cam rotates in synchronization with the crankshaft and ensures that the valve opens and closes at precisely the correct instants relative to the combustion cycle and the position of the piston. Figure 6.19 depicts an example of the type of computer simulations that mechanical engineers conduct when designing engine valvetrains.

Cam ---------------->

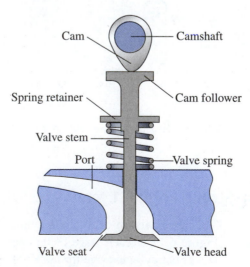

Cam   Camshaft

Spring retainer   Cam follower

Valve stem

Port   Valve spring

Valve seat   Valve head

FIGURE 6.18   Cam and valve mechanism for a four-stroke internal combustion engine.

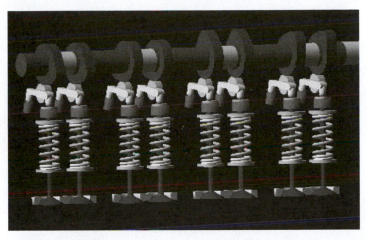

■ **FIGURE 6.19**   Mechanical engineers conduct computer simulations when designing valvetrains for multi-cylinder engines.
Source: Reprinted with permission of Mechanical Dynamics, Inc.

Otto cycle
- - - - - - - - - - - - - - →

Four-stroke engines operate on a continuous cycle (called the Otto cycle) that comprises four full strokes of the piston within the cylinder (or equivalently, two complete revolutions of the crankshaft). The engine's principle of operation is named after German engineer and inventor Nicolaus Otto (1832–1891), who is recognized for having developed the first practical design for liquid-fueled piston engines. Engineers use the abbreviation TDC (which stands for "top dead center") when referring to the point where the piston is at the top of the cylinder and the connecting rod and crank are in line with one another. Likewise, the term BDC stands for "bottom dead center," the orientation separated from TDC as the crankshaft rotates 180°. Referring to the sequence of piston positions shown in Figure 6.17, the four stages of the cycle's operation proceed as follows.

Top dead center
- - - - - - - - - - - - - - →

Bottom dead center
- - - - - - - - - - - - - - →

Intake Stroke   Just after TDC, the piston begins its motion down the cylinder. At this stage, the intake valve has already been opened, but the exhaust valve remains closed. As the piston moves downward, the cylinder's volume grows, and the pressure within the cylinder falls slightly below the outside atmospheric value. A fresh mixture of fuel and air flows into the cylinder. As the piston nears BDC, the intake valve closes, and the cylinder is now completely sealed.

Compression Stroke   The piston next travels upward in the cylinder so as to compress the fuel–air mixture. The ratio of volumes in the cylinder before and after this stroke takes place is called the engine *compression ratio*. Near the end of the stroke, the spark plug fires and ignites the high-pressure fuel

Compression ratio
- - - - - - - - - - - - - - →

and air. Combustion occurs rapidly at a nearly constant volume as the piston moves from a position slightly before TDC to a position slightly after. To visualize how the gas pressure within the cylinder changes during the Otto cycle, Figure 6.20 graphs cylinder pressure that was measured on a single-cylinder engine. The peak pressure for this engine was approximately 450 psi. For each pressure pulse, a portion of that rise in cylinder pressure was associated with the upward compressing motion of the piston in the cylinder, but the dominant factor was combustion following the point labeled "ignition" in Figure 6.20.

Power Stroke   With both valves remaining closed, the high-pressure gas in the cylinder forces the piston downward. The expanding gas performs work on the piston, which in turn drives the crankshaft. As shown in Figure 6.20, the pressure in the cylinder decreases quickly during the power stroke. As the stroke nears its end but while the pressure is still above the atmospheric value, a cam mechanism opens the exhaust valve. Some of the spent gas flows out from the cylinder through the exhaust port during the short stage, called *blow-down*, that occurs just before BDC.

Exhaust Stroke   After the piston passes through BDC, the cylinder remains full of exhaust gas at approximately atmospheric pressure. In the final stage of the Otto cycle and with the exhaust valve open, the piston moves upward toward TDC, forcing the spent gas from the cylinder. Near the end of the exhaust stroke and just before TDC is reached, the exhaust valve closes and the intake valve begins to open in preparation for the next repetition of the cycle.

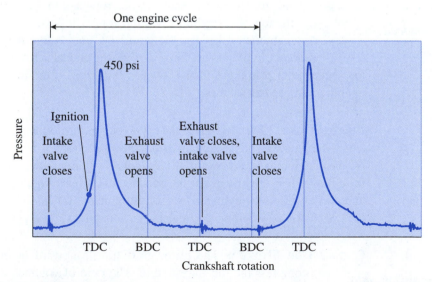

■ **FIGURE 6.20**   Measured pressure curve for a four-stroke engine operating at 900 rpm. There are 450 power strokes per minute, and the engine's compression ratio is $r = 6.3$.

To place this orchestration of valve and piston motions in perspective for the single-cylinder engine of Figure 6.20, consider that at 900 rpm, the crankshaft completes 15 revolutions each second and each of the four strokes occurs in only 33 ms. In fact, automobile engines often operate several times faster, and the short time intervals between valve openings and closings highlight the need for accurate timing and sequencing of the stages in the cycle. In the Otto cycle, only one stroke out of every four produces power, and in that sense, the crankshaft is driven only 25% of the time. The engine continues to rotate during the other three strokes because of momentum stored in the engine flywheel or, for a multicylinder engine, because of the overlapping power strokes from other cylinders.

On the basis of approximations in which the fuel–air mixture and exhaust are treated as ideal gases, the ideal efficiency of the Otto cycle is

$$\eta_{\text{otto}} = 1 - \frac{1}{r^{0.4}} \tag{6.21}$$

a value that is always less than the Carnot efficiency given in Equation 6.19. The quantity $r = V_{\text{BDC}}/V_{\text{TDC}}$ is the engine compression ratio, and it is the ratio of volumes in the cylinder with the piston at the bottom- and topmost positions. Aside from friction, incomplete combustion, and all other losses that are present in real engines, η_{otto} represents the maximum efficiency that can be obtained by an otherwise ideal engine operating on the four-stroke cycle. In the graph of Equation 6.21 shown in Figure 6.21, you can see how the efficiency of an engine can be raised by increasing its compression ratio. With a compression ratio of 7, for instance, the ideal efficiency is

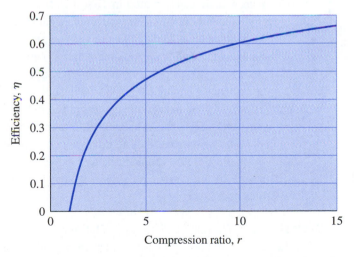

FIGURE 6.21   Efficiency of the four-stroke engine cycle as a function of compression ratio.

$\eta_{otto} = 1 - 1/7^{0.4} = 1 - 0.46 = 0.54 = 54\%$. By doubling the compression ratio, the efficiency can be increased to about 65%.

If engine efficiency increases with r, why not design engines that have compression ratios of 100, or 1000? The answer is that factors other than efficiency come into play when practical engines are built, and final designs always represent trade-offs and balance between competing considerations. In fact, early four-stroke automobile engines had compression ratios as low as $r \approx 2–4$. Although efficiency increases with the compression ratio, so does the gas pressure in the cylinder as the piston reaches TDC. Engines with high compression ratios must be stronger, heavier, and more bulky than others, or they must be made from higher strength and more expensive metals. As materials and designs have improved over the years, the compression ratios of commercial automobile engines have risen gradually to values of $r \approx 9–12$. The efficiency benefits associated with having higher compression ratios are balanced not only by the engine's weight and strength requirements but also by a problem known as self- or preignition. Should the compression ratio be too high or the octane rating of the fuel too low, the fuel–air mixture will spontaneously ignite as the piston nears the end of the compression stroke. Preignition creates pressure pulses that can damage the engine and produce the audible noise and vibration called *knocking* or *pinging*. In the United States prior to the 1970s, lead additives were incorporated in gasoline in order to reduce such problems. As those additives were phased out for health and environmental reasons, gasoline products and engine designs became more sophisticated to accommodate the changes in fuel composition.

Two-Stroke Engine Cycle

The second common type of internal combustion engine operates on the two-stroke or Clerk cycle, which was invented in 1880 by British engineer Dugald Clerk (1854–1932). Figure 6.22 depicts the cross section of a generic crankcase compression engine.

Transfer port

In contrast to its four-stroke cousin, the two-stroke engine has no valves, and hence no need for springs, camshafts, cams, or other valvetrain elements. Instead, a two-stroke engine has a passageway called the *transfer port* that connects the crankcase and cylinder, and fresh fuel–air mixture flows into the cylinder through that port. As the piston moves within the cylinder, it acts as the valve system by uncovering and sealing the exhaust port, intake port, and transfer port in sequence.

Clerk engine cycle

The two-stroke engine completes a full combustion cycle during each revolution of the crankshaft, and power is produced by it at every other stroke of the piston. As shown by the sequence of Figure 6.23, the Clerk cycle operates as follows.

Downstroke   The piston begins near TDC with a compressed mixture of fuel and air. After the spark plug fires, the piston is driven downward in its power stroke, and torque and rotation are transferred to the crankshaft. When the piston has moved partially down the cylinder, the exhaust port labeled in Figure 6.22 is uncovered. Because the gas in

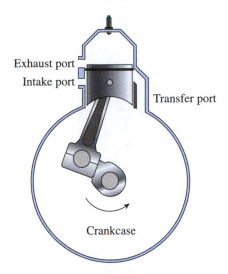

Exhaust port

Intake port

Transfer port

Crankcase

■ **FIGURE 6.22**   Cross section of a two-stroke internal combustion engine.

the cylinder is at an elevated pressure, it begins to vent outward through the open port. As discussed next for the upstroke stage, at this point in the cycle, a fresh charge of fuel and air is waiting in the crankcase, and it will be transferred into the cylinder for use in the engine's next cycle. The piston continues downward, and as the volume within the crankcase decreases, the pressure of the stored fuel–air mixture grows. Eventually, the transfer port becomes uncovered as the upper edge of the piston moves past it, and the fuel and air within the crankcase flow through the port and fill the cylinder. During this process, the piston continues to block the intake port.

Upstroke   Once the piston passes BDC, most of the exhaust gas has been expelled from the cylinder. As the piston moves upward, the transfer port and exhaust port each become covered, and the pressure in the crankcase falls because it is expanding. The intake port is uncovered as the lower edge of the piston moves past it, and fresh fuel and air flow into the crankcase. That mixture, in turn, is stored for use during the next combustion cycle. Slightly before the piston reaches TDC, the spark plug fires, and the power cycle begins anew.

Two- and four-stroke engines each have their own advantages. Relative to their four-stroke counterparts, two-stroke engines are simpler and less expensive. Mechanical engineers appreciate the saying "keep it simple, stupid," and because two-stroke engines have few moving parts, there is little to go wrong with them. On the other hand, the intake, compression, power, and exhaust stages of operation in a two-stroke engine are not as well separated as they are in a four-stroke engine. The (spent) exhaust and (new) fuel–air gases mix to an unavoidable extent as the new charge flows from the crankcase into the cylinder. For that reason, some amount of unburned fuel is present in a two-stroke engine's exhaust, contributing to environmental pollution and reduced fuel economy. Because the crankcase is needed to store fuel–air charges, it cannot be

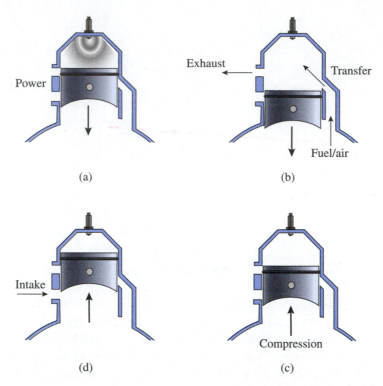

■ **FIGURE 6.23** Sequence of stages in the operation of a two-stroke internal combustion engine.

used as an oil sump, as is the case in a four-stroke engine. Lubrication in a two-stroke is instead provided by oil that has been premixed with the fuel, further contributing to emissions problems.

6.7 CASE STUDY #2: ELECTRICAL POWER GENERATION

The vast majority of electrical energy used in the United States is produced by power plants that employ a thermal and energy cycle based on water and steam. Other contributing sources include the renewable technologies of wind, solar, and hydroelectric power. To place the technology of electrical power into perspective, in 1999, the United States had a generation capability of 786 GW, and 3.7 trillion kW·h of energy were produced. The energy unit of kW·h (which is the product of units for power and time) is the amount produced in 1 hour by a 1-kilowatt source; a 100-W lightbulb, for instance, consumes 0.1 kW·h of energy each hour. As you can see from Table 6.9 for various technologies, fossil fuel plants account for approximately 73% of all generation capability, and nuclear power plants produce some 12% of electricity in the United States.

TABLE 6.9 Statistics for Electrical Power Generation in 1999. "Other" Sources Include Geothermal, Solar, and Wind Technologies.

Power Plant Type	Capacity (GW)	Contribution (%)
Coal fired	313	40
Petroleum fired	86	11
Gas fired	171	22
Nuclear powered	97	12
Hydroelectric	98	13
Other	20	2
Total	786	100

Source: United States Department of Energy, Energy Information Administration.

In the macroscopic view, a fossil fuel power plant receives fuel and air as its inputs. In turn, the plant produces electricity as its primary output, along with two side effects that are released into the environment: the waste products of combustion, and residual heat. A power plant requires large quantities of cooling water for the portion of its thermal and energy cycle in Figure 6.12 where waste heat is rejected. Electrical power plants are often located near large bodies of water, or they will have cooling towers, in order to release that heat. Waste heat is sometimes called *thermal pollution*, and it can disturb both the habitat of wildlife and the growth of vegetation.

Primary loop

- - - - - - - - - - - - - - ➤

Rankine cycle

- - - - - - - - - - - - - - ➤

As shown in Figure 6.24, the cycle for a fossil fuel power plant is a loop that comprises the boiler, turbine, condenser, and pump. Continuously converting water between a liquid state and a vapor steam, the power plant operates on what is called the *Rankine cycle*, named in honor of the Scottish engineer and physicist William John Rankine (1820–1872). The water–steam physically moves energy from one location in the power plant (for instance, the boiler) to another (the turbines). Some 85% of the electrical power in the United States is produced through this process in one variation or another.

Pump

- - - - - - - - - - - - - - ➤

Boiler

- - - - - - - - - - - - - - ➤

Beginning our description of the power plant at the pump, liquid water is pressurized and flows to the boiler. A relatively small amount of mechanical work, W_{in} is supplied to the power plant as a whole to drive the pump. As fuel burns, the heat, Q_{in} released is transferred to the water, and steam is produced at high temperature and pressure. The combustion gases heat a network of tubes in the boiler called a *heat exchanger*, and the water turns to steam while being pumped through the tubes. The steam then flows through pipes to the turbines, where it expands and performs mechanical work. Each stage of the turbine

Turbine

- - - - - - - - - - - - - - ➤

is analogous to a water wheel, and as jets of steam strike the blades, the shaft of the turbine is forced to rotate. The turbine is connected to a generator that ultimately produces the electricity leaving the power plant as W_{out}. The spent

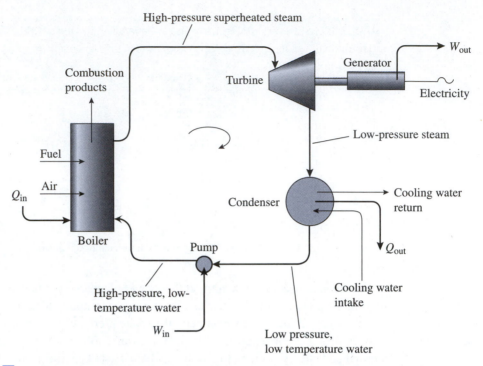

FIGURE 6.24 Schematic of the cycle used for electrical power generation.

Condenser
- - - - - - - - - - - - - - →

low-pressure steam leaves the turbine and next enters the condenser, where the quantity of heat Q_{out} is transferred to the environment. The steam condenses down to low-pressure and low temperature water so that it can be pumped and recirculated through the system. Water can be more easily and efficiently pumped than the compressible steam vapor. It is important to note that there is no mixing between water in the primary loop and cooling water from a lake, river, or ocean. The cycle repeats as the water exits the pump and returns to the boiler.

Following the first law in Equation 6.10, the energy balance for the power plant is $Q_{in} + W_{in} = Q_{out} + W_{out}$. In turn, the real efficiency is measured in terms of the useful output W_{out} and the required inputs Q_{in} and W_{in} according to

$$\eta = \frac{W_{out}}{Q_{in} + W_{in}} = \frac{Q_{in} + W_{in} - Q_{out}}{Q_{in} + W_{in}} = 1 - \frac{Q_{out}}{Q_{in} + W_{in}}$$

$$\approx 1 - \frac{Q_{out}}{Q_{in}} \tag{6.22}$$

In the last portion of this equation, we have made the simplification that the work supplied to the pump is very small when compared to the plant's heat input and output. The overall efficiency of real power plants from fuel to electricity is in the range of $\eta = 35\text{–}40\%$.

EXAMPLE 6.16

A 1000-MW power plant burns coal and produces electricity. The boiler operates at 550°C, and the condenser releases waste heat at 25°C. By using the Carnot efficiency as an estimate, calculate the plant's daily fuel consumption.

SOLUTION

From this information, we don't know the plant's real efficiency η, but it must be less than the ideal efficiency expressed by Equation 6.19. We will use that value as an estimate. On the absolute scale, the high- and low-temperature heat reservoirs operate at $T_h = (273 + 550)°K = 823°K$ and $T_l = (25 + 273)°K = 298°K$. The ideal efficiency is therefore

$$\eta_c = 1 - \frac{T_l}{T_h} = 1 - \frac{298°K}{823°K} = 0.64 = 64\%$$

The plant's output of "what we get" relative to "what we paid for" is

$$W = \eta Q_h$$

or $Q_h = W/\eta$. In 1 s the plant produces 1000 MJ, so that $(1000 \text{ MJ})/0.64 = 1568 \text{ MJ}$ of heat must be provided to the boiler. In Table 6.3, the heat of combustion for coal is listed as 2.8×10^4 kJ/kg. The amount of fuel necessary to produce 1568 MJ is

$$(1568 \text{ MJ})(1000 \text{ kJ/MJ}) = m(2.8 \times 10^4 \text{ kJ/kg})$$

Each second, the plant must burn $m = 56$ kg of coal, and in 1 day, 4.8 million kg of coal are consumed.

It is also instructive to consider the effects of pollution, because a key problem faced by fossil fuel power plants is the emission of sulfur as a combustion by-product. Sulfur reacts with rainwater in the atmosphere to produce sulfur dioxide, and in turn, sulfuric acid, which is the main ingredient of acid rain. Coal typically contains about 0.5–4% sulfur by weight. If the power plant burns 2% sulfur coal, then in the absence of any effort to clean the exhaust gases, each day some $(0.02)(4.8 \times 10^6 \text{ kg}) = 9.6 \times 10^4$ kg of sulfur will have been released into the atmosphere. In fact, power plants in the United States are equipped with scrubber systems that remove a large fraction of the sulfur from the exhaust stack. If the system should be 95% efficient, some $(1 - 0.95)(9.6 \times 10^4 \text{ kg}) = 4800$ kg of sulfur would still be released into the air each day. You should bear in mind that not all countries use such systems for cleaning exhaust gas, and the large-scale pollution effects of fossil fuel consumption can indeed be high. Mechanical engineers have an active role to play in the public policy arena by balancing environmental issues with the need for an abundant and inexpensive supply of electricity (Figure 6.25). ■

In a nuclear power plant, a reactor serves as the heat source, and fuel rods comprised of a radioactive material replace the fossil fuels coal, oil, or natural gas. Reactors operate on the principle of nuclear fission, by which the structure of matter is altered

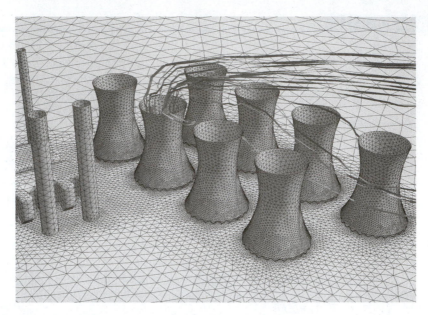

at the atomic level. Energy is released in large quantity as the nucleus of an atom is split. Fuels consumed in this manner have an enormous energy density, and 1 g of the uranium isotope U-235 is capable of producing the same amount of heat as some 3000 kg of coal.

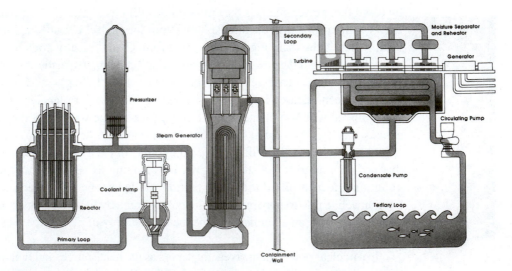

The layout of a generic nuclear power plant is shown in Figure 6.26. It differs from the fossil fuel plant of Figure 6.24 in that the nuclear plant has two distinct internal loops. Water flowing in the primary loop comes into direct contact with the nuclear fuel, and it transfers heat from the reactor to the steam generator. For safety reasons, the primary loop and the water within it do not pass outside a hardened containment wall. The steam generator functions as a means to transfer heat from the water in the primary loop to the water in the secondary loop and to otherwise keep the two loops physically isolated. The water–steam of the secondary loop drives the turbines and generators in a conventional steam cycle. At the outermost level, a third or tertiary loop draws water for cooling purposes in the condenser. It is important to note that the water supplies within these three loops do not mix. You can gain an appreciation for the internal construction details of the reactor and steam generator from the cross-sectional drawings of Figures 6.27 and 6.28.

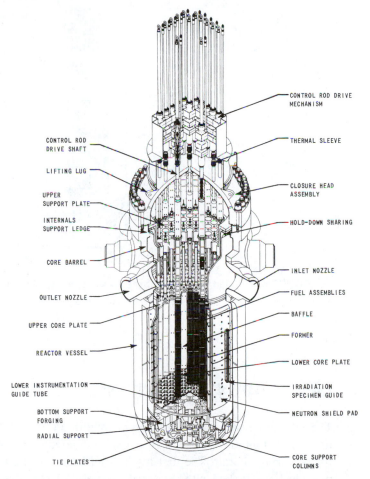

■ **FIGURE 6.27**   Cutaway view of the reactor in a nuclear power plant.
Source: Reprinted with permission of Westinghouse Electric Company.

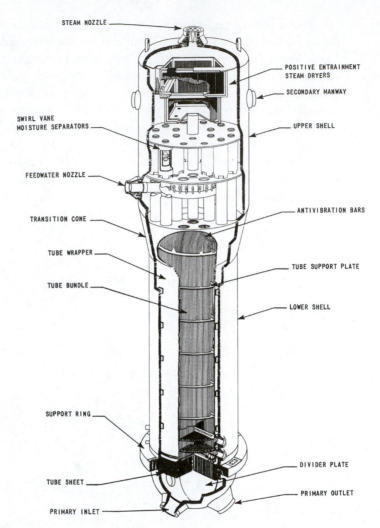

STEAM NOZZLE

POSITIVE ENTRAINMENT
STEAM DRYERS

SECONDARY MANWAY

SWIRL VANE
MOISTURE SEPARATORS

UPPER SHELL

FEEDWATER NOZZLE

ANTIVIBRATION BARS

TRANSITION CONE

TUBE WRAPPER

TUBE SUPPORT PLATE

TUBE BUNDLE

LOWER SHELL

SUPPORT RING

DIVIDER PLATE

TUBE SHEET

PRIMARY OUTLET

PRIMARY INLET

■ **FIGURE 6.28** Cutaway view of the steam generator in a nuclear power plant.
Source: Reprinted with permission of Westinghouse Electric Company.

6.8 CASE STUDY #3: JET ENGINES

The jet engine was invented by Frank Whittle (1907–1996) in 1929, but it was not until the height of World War II that substantial attention was given to it as a means for powering aircraft. The first flight of an experimental jet aircraft in England took place in 1941. In the United States, the General Electric Corporation further developed Whittle's designs and had prototyped three jet aircraft by the war's end. Although others had experimented with jet engines as a means of aircraft propulsion, Whittle's contribution was the idea to insert a turbine downstream of the combustion chamber so that the turbine could drive the engine's compressor (Figures 6.29, 6.30). In doing so, it was possible to obtain a higher compression ratio, enabling the engine to be both powerful and lightweight.

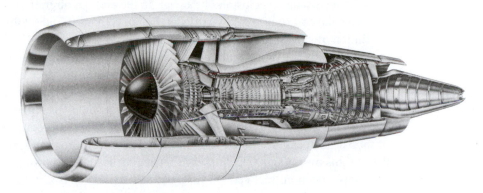

■ **FIGURE 6.29**   Cutaway view of a jet engine.
Source: Reprinted with permission of GE Aircraft Engines.

■ **FIGURE 6.30**   Mechanical engineers use computer-aided engineering software to calculate the temperature of turbine blades in a jet engine.
Source: Reprinted with permission of Fluent Inc.

A parallel effort had been underway in Germany by an engineering student named Hans von Ohain (1911–1998), who was only 22 years old when he began to develop his ideas for jet engine propulsion. Von Ohain's designs were evidently conceived some time after Whittle's, but for various reasons, jet engines moved to production more rapidly in Germany. In fact, German pilots first flew turbojet-powered planes in 1939, approximately 2 years before the British, and some 1000 jets had been manufactured in Germany by the conclusion of the war. In the end, jet aircraft did not substantially influence the course of World War II, but their development did set the stage for a rapid revolution of aviation in the 1950s and 1960s.

Essentially, all long-haul commercial aircraft flights are today powered by jet engines. Similar systems called *gas turbines* are used to produce electrical power and to drive ships, military tanks, and other large vehicles. To place the scale of jet engine performance in perspective, the engines that power some Boeing 777 aircraft have an inlet diameter of just over 10 ft, and they are each rated at producing over 76,000 lb of thrust. Such engines power long-range cruising at speeds of Mach 0.84 for a plane that weighs over 500,000 lb at takeoff. Jet engine design for power, thrust, low weight, materials selection, fuel economy, safety, and reliability is a remarkable mechanical engineering achievement, and it involves practical applications, extreme environmental conditions, and high technologies.

Layouts for the two main types of jet engines are compared in Figure 6.31. In a turbojet engine, air is drawn into the engine's front, is compressed, has fuel added to it, is ignited and burned, and then expands and exhausts through a nozzle located at the rear of the engine. The high velocity of the gas as it exits the nozzle provides the aircraft with forward thrust. That sequence of stages occurs in a module called the engine's "core," and a schematic of its internal construction is shown in Figure 6.32. A turbofan

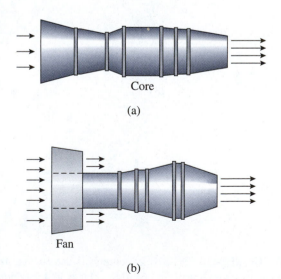

FIGURE 6.31   Paths of airflow in (a) turbojet and (b) turbofan engines.

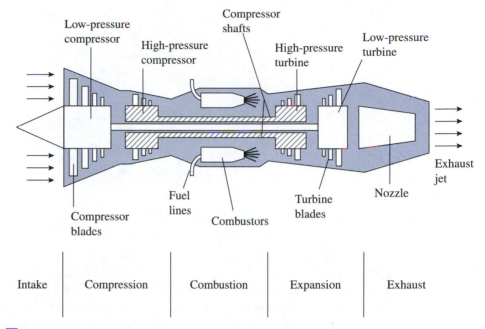

FIGURE 6.32   Cross-sectional view of the core in a jet engine.

engine, shown in Figure 6.31(b), is also formed around the core, but it incorporates large fan blades that are directly driven by the engine. While some fraction of thrust is provided by the exhaust gas as it exits the nozzle, the fan provides a significant amount of additional thrust.

Jet engines follow the thermal and energy cycle illustrated in the block diagram of Figure 6.33. Air enters the front of the engine and is compressed in multiple stages. The compressor is powered by a shaft that extends down the engine centerline to the

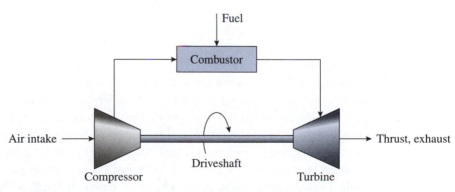

FIGURE 6.33   Schematic of the power cycle for a jet engine.

■ **FIGURE 6.34** Turbine stage in a jet engine.

turbine. Near the center of the engine, the compressed fuel–air mixture is ignited and burns in the combustion chamber. The high-pressure and high-temperature gas expands, causing the turbine to rotate. As shown in Figures 6.30 and Figure 6.34, each stage of the turbine is similar in appearance to a fan and comprises many short, specially shaped, blades. The shafts that provide power to the compressor extend forward from the turbine along the length of the engine. The turbine produces just enough power to drive the compressor, and the energy remaining in the combustion gas provides the plane with its thrust.

SUMMARY

In this chapter, we introduced several principles behind the thermal and energy systems encountered in mechanical engineering. Internal combustion engines, electrical power plants, and jet aircraft engines involve the conversion of heat to work, and each figured prominently in Chapter 1 during our discussion of the mechanical engineering profession's top ten achievements. The different forms that energy can take and the methods by which energy can be efficiently exchanged from one form to another are, in short, central to mechanical engineering.

The main quantities introduced in this chapter, common symbols representing them, and their units are summarized in Table 6.10. Heat engines are often powered by burning a fuel, and the heat of combustion, H, represents the quantity of heat released during that combustion process. In accordance with the first law, the heat can flow into a material and cause its temperature to rise by an amount depending on its specific heat c, or the heat can be transformed into mechanical work. We view heat as energy in transit through the modes of conduction, convection, and radiation. Mechanical engineers design engines that convert heat into work with a view toward increasing their efficiency. In an ideal or Carnot heat engine, the efficiency is limited by the temperatures at which heat is supplied to the engine and released to its environment. In the final three sections of this chapter, we described the operation of four-stroke and two-stroke internal combustion engines, fossil fuel and nuclear power plants, and jet engines in order to introduce some practical hardware behind the thermal and energy systems of mechanical engineering.

TABLE 6.10 Quantities, Symbols, and Units That Arise in the Analysis of Thermal and Energy Systems

Quantity	Conventional Symbols	Conventional Units	
		USCS	SI
Energy	ΔU	ft·lb, Btu	J, kW·h
Work	$\Delta W, W$	ft·lb	J
Heat	$\Delta Q, Q$	Btu	J
Average power	P_{avg}	ft·lb/s, hp	W
Heat of combustion	H	Btu/slug	MJ/kg
Specific heat	c	Btu/(slug·°F)	kJ/(kg·°C)
Thermal conductivity	κ	(Btu/h)/(ft·°F)	W/(m·°C)
Temperature change	ΔT	°F	°C
Absolute temperature	T_h, T_l	°R	°K
Efficiency	η	—	—
Carnot efficiency	η_c	—	—
Otto cycle efficiency	η_{otto}	—	—
Compression ratio	r	—	—

SELF-STUDY AND REVIEW

1. How are gravitational and elastic potential energy, and kinetic energy, calculated?
2. What is the difference between work and power? What are their conventional units in the SI and USCS?
3. What is the "heat of combustion" of a fuel?
4. What is the "specific heat" of a material?
5. Explain how heat is transferred through the processes of conduction, convection, and radiation.
6. Define the term *thermal conductivity*.
7. What is a heat engine, and how is its efficiency calculated?
8. What are the differences between actual and Carnot efficiencies?
9. How are absolute temperatures determined on the Rankine and Kelvin scales?
10. Sketch the cross sections of four-stroke and two-stroke engines, and explain how they work.
11. What are the relative advantages and disadvantages of four-stroke and two-stroke engines?

12. Sketch a block diagram of an electric power plant, and explain how it works.
13. Sketch a diagram of a jet engine, and explain how it works.

PROBLEMS

1. Beginning from their definitions, determine the conversion factor between ft lb and kW·h in Table 6.1.

2. In the movie *Back to the Future*, Doc Brown and the young Marty McFly need 1.21 GW of power for their time machine. (a) Convert that power requirement to horsepower. (b) If a stock DeLorean sports car produces 145 hp, how many times more power does the time machine need?

3. For the two automobiles of Problem 36 in Chapter 4, how much power must the engines produce just to overcome air drag at 60 mph?

4. During processing in a mill, a 750-lb steel casting at 800°F is quenched by plunging it into a 500-gal oil bath, which is initially at a temperature of 100°F. After the casting cools and the oil bath warms, what is the final temperature of the two? The weight per unit volume of the oil is 7.5 lb/gal.

5. The interior contents and materials of a small building weigh 50,000 lb, and together they have an average specific heat of 8 Btu/(slug·°F). Neglecting any inefficiency in the furnace, what amount of natural gas must be burned to raise the building temperature from freezing to 70°F?

6. Give two examples each of engineering applications where heat would be transferred primarily through conduction, convection, and radiation.

7. A hollow square box is made from 1-ft² sheets of a prototype insulating material that is 1-in. thick. Engineers are performing a test to measure the new material's thermal conductivity. A 100-W electrical heater is placed inside the box. Over time, thermocouples attached to the box show that the interior and exterior surfaces of one face have reached the constant temperatures of 150°F and 90°F. What is the thermal conductivity? Express your result in both the SI and USCS.

8. A welding rod with $\kappa = 30(\text{Btu/h})/(\text{ft·°F})$ is 20 cm long and has a diameter of 4 mm. The two ends of the rod are held at 500°C and 50°C. (a) In the units of Btu and J, how much heat flows along the rod each second? (b) What is the temperature of the welding rod at its midpoint?

9. A 2500-lb automobile comes to a complete stop from 65 mph. If 60% of the braking capacity is provided by the front disk brake rotors, determine their temperature rise. Each of the two cast-iron rotors weighs 15 lb and has a specific heat of 4.5 Btu/(slug·°F).

10. A small hydroelectric power plant operates with 500 gal of water passing through the system each second. The water falls through a vertical distance of 150 ft from a reservoir to the turbines. Calculate the power output, and express it in the units of both hp and kW. The density of water is listed in Table 4.3.

11. Neglecting the presence of friction, air drag, and other inefficiencies, how much gasoline is consumed when a 1300-kg automobile accelerates from rest to 80 km/h? Express your answer in the units of mL. The density of gasoline is listed in Table 4.3.

12. A group of miners has become trapped 240 ft underground when a section of the coal seam they were working collapsed into an adjacent, but abandoned, mine that was not shown on their map. The area is becoming flooded with water, and the miners have huddled in an air

pocket at the end of a passageway. As the first step in the rescue operation, holes have been drilled into the mine to provide the miners with warm fresh air and to pump out the underground water. Neglecting friction in the pipes and inefficiency of the pumps themselves, what average power is required to remove 20,000 gal of water from the mine each minute? Express your answer in the units of horsepower. The density of water is listed in Table 4.3.

13. (a) Use the principle of dimensional consistency to show that when Bernoulli's equation is written in the form

$$p + \frac{1}{2}\rho v^2 + \rho g h = \text{constant}$$

each term has the unit of pressure. (b) When the equation is written

$$\frac{p}{\rho g} + \frac{v^2}{2g} + h = \text{constant}$$

show that each term has the unit of length.

14. Water flows through a circular pipe having a diameter reduction from 1 in. to $\frac{1}{2}$ in. If the velocity of the water just upstream of the constriction is 4 ft/s, what are (a) the pressure drop across the constriction and (b) the velocity of the water downstream from it? The density of water is listed in Table 4.3.

15. Prepare a calibration graph that an engineer can use to relate airspeed and pressure difference in a pitot tube. The graph will be used during low-speed wind tunnel testing so that a measured pressure difference can be quickly converted to airspeed. Use the units of psi for pressure, and mph for airspeed on the graph's axes.

16. Geothermal energy systems extract heat that is stored below the Earth's crust. For every 300 ft below ground, the temperature of groundwater increases by 5°F. Heat can be brought to the surface by steam or hot water

to warm homes and buildings, and it can also be processed by a heat engine to produce mechanical work or electricity. Using the Carnot efficiency as an approximation, calculate the output of a geothermal power plant that processes 50 lb/s of groundwater at 180°F and discharges it on the surface at 70°F.

17. A heat engine idealized as operating on the Carnot cycle is supplied with heat at the boiling point of water (212°F) and rejects heat at the freezing point of water (32°F). If the engine produces 100 hp of mechanical work, calculate in the units of Btu the amount of heat that must be supplied to the engine each hour.

18. A person can blink an eye in approximately 7 ms. At what speed (in revolutions per minute) would a four-stroke engine be operating if its power stroke took place literally in the "blink of an eye"? Is that a reasonable speed for an automobile engine?

19. A four-stroke gasoline engine has a compression ratio of 8, and it produces an output of 35 kW. What is the maximum possible efficiency of the engine? Using the heat of combustion for gasoline in Table 6.3, estimate the engine's rate of fuel consumption. Express your answer in the units of liters per hour. The density of gasoline is listed in Table 4.3.

20. An automobile engine with a compression ratio of 10 produces 30 hp while being driven at 50 mph on a level highway. In those circumstances, the engine power is used to overcome air drag, rolling resistance between the tires and the road, and friction within the drivetrain. Estimate the fuel economy rating of the vehicle in the units of miles per gallon. The density of gasoline is listed in Table 4.3.

21. An electrical power plant produces an output of 200 MW. As coal is burned, heat at 600°C is supplied to the boiler at the rate of

700 MW. The condenser releases 520 MW of heat into a 20°C lake that is adjacent to the plant. (a) Determine the power consumed by the pump, and (b) compare the real efficiency of the power plant with the ideal Carnot efficiency.

22. An electrical power plant produces an output of 750 MW. The boiler operates at 950°F, and the condenser rejects heat into a river at 50°F. Using the Carnot efficiency as an estimate and neglecting the small amount of power drawn by the pump, calculate the rates at which heat is (a) supplied to the boiler and (b) spent into the river.

23. For the plant of Problem 22, 10,000 gal/s of water flow in the river adjacent to the power plant. Considering the heat transferred from the power plant to the river, and the specific heat of water, by what amount does the temperature of the river rise as it passes the power plant? The density and specific heat of water are listed in Tables 4.3 and 6.5.

7 Motion of Machinery

7.1 OVERVIEW

In this chapter, we turn our attention to the motion of machinery as a fifth element of mechanical engineering. The analysis and design of machinery encompasses several topics that we have encountered in previous chapters: gears, bearings, and other machine components (Chapter 3), force systems (Chapter 4), and energy and power (Chapter 6). In Chapter 3, for instance, we described various types of gears and the circumstances in which one would be selected over another. At the next level of sophistication, sets of gears can be assembled into geartrains and transmissions for the purposes of transmitting power, changing the rotation speed of a shaft, or modifying the torque that is applied to a shaft (Figure 7.1). To those ends, in the first portion of this chapter, we develop some of the methods used by mechanical engineers when they analyze the geartrain designs known as simple, compound, and planetary.

Aside from gears alone, machinery can also include links, shafts, bearings, springs, cams, and other building-block components in order to transform one type of motion into another (Figure 7.2). An example that we encountered in Figure 6.14 is the mechanism for converting the back-and-forth motion of the piston in an internal combustion engine into the crankshaft's rotation. The shapes of cams that open and close the intake and exhaust ports within four-stroke engines (Figure 6.18) are likewise carefully designed. Together with the thermal and energy characteristics described in Chapter 6, mechanical engineers evaluate the position, velocity, and acceleration of machine components and the forces that cause them to move. In Section 7.6, we will explore the motion of machinery in the context of the piston, connecting rod, and crankshaft linkage of an engine or compressor.

In the following sections, we develop some of the tools that are necessary to understand how machinery functions with a view toward practical geartrains and mechanisms. Later, Chapter 8 includes a case study of the design of the first automobile automatic transmission, a complex system comprising geartrains, clutches, brakes, and hydraulic

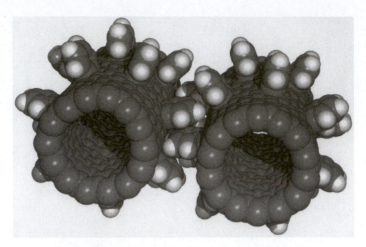

■ **FIGURE 7.1**   The field of nanotechnology is concerned, in part, with the construction of molecular-sized mechanical devices. Researchers are investigating nanometer-sized "gears" that are formed by attaching benzene molecules to the outside of pipes made of carbon atoms. A laser serves as the "motor" to drive the gearset.
Source: Reprinted with permission of NASA.

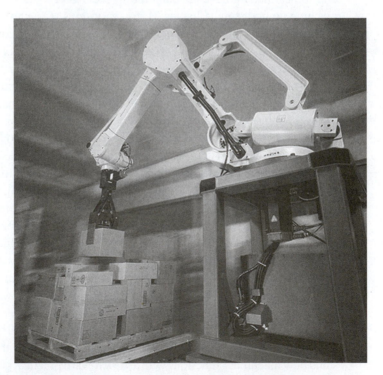

■ **FIGURE 7.2**   Mechanical engineers design the links and joints that form this robot's manipulator arm in order to precisely move and place cargo in a warehouse.
Source: Reprinted with permission of FANUC Robotics North America Inc.

controls. The case study summarizes the elegant construction of the transmission, and it also places in context the major contribution of the transmission for consumers. After completing this chapter, you should be able to:

- Understand the different choices of units for angular velocity, and make appropriate conversions between them.
- Calculate the speeds of gears, the torques acting on their shafts, and the power transferred in simple and compound geartrains.
- Determine the overall velocity and torque ratios for a geartrain.
- Sketch a planetary geartrain and explain how it operates.
- Relate the motions of the piston and crankshaft in an internal combustion engine or compressor.

7.2 ROTATIONAL MOTION

Rotational Velocity

When a gear or any other object rotates, each point on it moves in a circle about the center of rotation (Figure 7.3). The straight link in Figure 7.4, for instance, turns in a bearing about the center of its shaft. All points on the link move along concentric circles, each having center O, as the angle θ (the lowercase Greek character theta) increases. The velocity of any point P is determined by its change in position as the rotation angle changes. During the time interval Δt, the link moves from the initial angle θ to a final angle $\theta + \Delta\theta$; the latter position is shown by the dashed lines in Figure 7.4(a). As P rotates along a circle of radius r, the distance that it moves is the arc length $\Delta s = r\,\Delta\theta$. For dimensional consistency, since Δs has the units of length, the angle $\Delta\theta$ is expressed in the units of radians (and not degrees). The velocity of point P is the distance traveled

Angular velocity
-------------------➤

during the time interval, or $v_p = \Delta s / \Delta t = r(\Delta\theta / \Delta t)$. In standard form, we define ω (the lowercase Greek character omega) as the rotational or angular velocity of the link, and the velocity is

$$v_p = r\omega \tag{7.1}$$

The appropriate units for ω are chosen based on context. When a mechanical engineer refers to the speed of an engine, shaft, or gear, it is customary to use the units of revolutions per minute (rpm), which a tachometer would measure. If the rotation speed is very high, the units of revolutions per second (rps) might be used instead. It would be inaccurate, however, to express ω in either rpm or rps when Equation 7.1 is applied. To understand why, recall that in the derivation for v_p, ω was defined by the angle change $\Delta\theta$, which must be measured in radians in order for the arc length Δs to be properly calculated. Whenever you apply Equation 7.1, ω must be converted to the units of radians per unit time (for instance, rad/s). If ω is known in rad/s and r is given in millimeters, then v_p has the units of millimeters per second. Table 7.1 lists the conversion factors between four common choices of units for

■ **FIGURE 7.3** Suspended above the bay of the space shuttle orbiter, the Hubble Space Telescope is maneuvered into sunlight during a servicing mission. The links of the robotic manipulator arm each rotate about their support joints.
Source: Reprinted with permission of NASA.

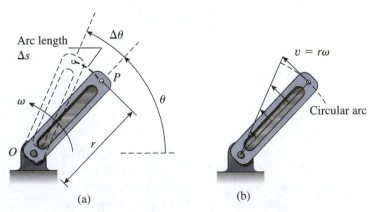

■ **FIGURE 7.4** (a) Rotational motion of a straight link. (b) Each point moves in a circle, and the velocity increases with radius from the pivot point.

TABLE 7.1   Conversion Factors Between Various Units for Angular Velocity

rpm	rps	deg/s	rad/s
1	0.0167	6	0.1047
60	1	360	6.283
0.1667	0.0028	1	0.0175
9.549	0.1592	57.30	1

angular velocity: rpm (revolutions per minute), rps (revolutions per second), degrees per second, and radians per second. For instance, by using Table 7.1, we see in the first row that 1 rpm = 0.0167 rps = 6 deg/s = 0.1047 rad/s, and from the fourth row that 1 rad/s = 0.1592 rps.

EXAMPLE 7.1

Determine the conversion factor between rpm and rad/s in Table 7.1.

SOLUTION

We convert one revolution to radians, and one minute to seconds, according to

$$1 \text{ rpm} = 1\frac{\text{rev}}{\text{min}} = \left(1\frac{\text{rev}}{\text{min}}\right)\left(\frac{1}{60}\frac{\text{min}}{\text{s}}\right)\left(2\pi\frac{\text{rad}}{\text{rev}}\right) = 0.1047 \text{ rad/s}$$

EXAMPLE 7.2

The fan shown in Figure 7.5 spins at 1800 rpm and cools a rack of electronic circuit boards. (a) Express the angular velocity in the units rps, deg/s, and rad/s. (b) Determine the velocity of point A at the tip of the 4-cm fan blade.

SOLUTION

(a) Referring to the conversion factors in Table 7.1, we express the angular velocity in each choice of units as follows:

$$\omega = (1800 \text{ rpm})\left(0.0167\frac{\text{rps}}{\text{rpm}}\right) = 30 \text{ rps}$$

$$\omega = (1800 \text{ rpm})\left(6\frac{\text{deg/s}}{\text{rpm}}\right) = 1.08 \times 10^4 \text{ deg/s}$$

$$\omega = (1800 \text{ rpm})\left(0.1047\frac{\text{rad/s}}{\text{rpm}}\right) = 188.5 \text{ rad/s}$$

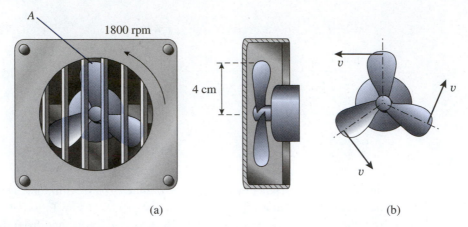

(a)   (b)

■ **FIGURE 7.5** The small cooling fan powered by an electric motor in Example 7.2.

(b) In Figure 7.5(b), the velocity of each blade tip is perpendicular to the radius line drawn to the shaft's center. By using Equation 7.1,

$$v_A = (4\text{ cm})(188.5\text{ rad/s}) = 754\text{ cm/s} = 7.54\text{ m/s}$$ ■

Rotational Work and Power

In addition to specifying the speeds at which shafts rotate, mechanical engineers also determine the amount of power that machinery draws, transfers, or produces. As expressed by Equation 6.5, power is defined as the rate at which work is performed. The mechanical work itself can be associated with forces moving through a distance (which we considered in Chapter 6) or, by analogy, with torques rotating through an angle. The definition of a torque T in Chapter 4 as a moment that acts about the axis of a shaft makes the latter case particularly relevant for machinery, geartrains, and mechanisms.

Figure 7.6 illustrates the work of a torque in the context of a motor and gear. The motor applies torque to the gear, and work is performed during rotation. Analogous to Equation 6.4, torque work is calculated from

$$\Delta W = T\,\Delta\theta \tag{7.2}$$

Work of a torque
- - - - - - - - - - - - - - ➤

where the angle $\Delta\theta$ has the unit of radians. As is the case for the work of a force, the sign of ΔW depends on whether the torque tends to reinforce the rotation (positive sign) or oppose it (negative sign). Engineering units for work are summarized in Table 6.1.

Mechanical power is the rate at which the work of a force or torque is performed.

Instantaneous power
- - - - - - - - - - - - - - ➤

In machinery applications, power is generally expressed in the units of kW in the SI and hp in the USCS (Table 6.2). We defined the average power in Equation 6.5, but the instantaneous power

$$P = Fv \quad \text{or} \quad P = T\omega \tag{7.3}$$

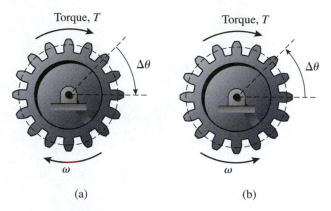

Torque, T   Torque, T

$\Delta\theta$ $\Delta\theta$

ω ω

(a)   (b)

FIGURE 7.6 Work performed by a torque that acts on a rotating gear. (a) The torque supplied by a motor reinforces the rotation ($\Delta W > 0$). (b) The torque supplied by the mechanical load opposes the rotation ($\Delta W < 0$).

is often more useful when analyzing machinery. In short, instantaneous power is the product of force and velocity, or torque and angular velocity.

EXAMPLE 7.3

At full throttle, a four-cylinder automobile engine produces 149 ft·lb of torque at 3600 rpm. In the units of hp, what is the power output of the engine?

SOLUTION

We apply Equation 7.3 to calculate the instantaneous power. The engine's speed must first be converted to the consistent units of rad/s

$$\omega = (3600 \text{ rpm}) \left(0.1047 \; \frac{\text{rad/s}}{\text{rpm}} \right) = 377 \text{ rad/s}$$

The power developed by the engine is then

$$P = (149 \text{ ft·lb})(377 \text{ rad/s}) = 5.617 \times 10^4 \text{ ft·lb/s} = 102 \text{ hp}$$

In the last step, we used the conversion factor 1 hp = 550 ft·lb/s from Table 6.2. ∎

7.3 SPEED, TORQUE, AND POWER IN GEARSETS

Gears transmit rotation, torque, and power between shafts. A gearset is a pair of two gears that mesh with one another, and it forms the basic building block of larger-scale systems. In the multiple-speed transmission of Figure 7.7, for instance, the shift fork is used to slide a spur gear along the splined shaft in order to change the speed of the output shaft. In this section, we examine the speed, torque, and power characteristics of

Shift fork

Splined shaft

Spur gears

■ FIGURE 7.7 The shift fork in the transmission is used to slide a gear along the upper shaft so that the speed of the output shaft changes.

two meshing gears, and in later sections, we apply those results to simple, compound, and planetary geartrains.

In Figure 7.8(a), the smaller of the two gears is called the *pinion* (denoted p), and the larger is called the gear (g). The pitch radii of the pinion and gear are denoted by r_p and r_g, respectively. In order that the pinion and gear mesh smoothly, they must be manufactured with the same distance between teeth, as measured by either the diametral pitch or module. In both Equations 3.1 and 3.2, the number of teeth N is proportional to the radius r. In the gearset of Figure 7.8, and in the simple, compound, and planetary geartrains that we will consider in the following sections, the gear teeth have ideal profiles. In that way, the fundamental property of gearsets is satisfied, and the meshing process between the pinion and gear is equivalent to two cylinders rolling on one another.

Speed

As the pinion rotates, the speed of a point on its pitch circle is $v_p = r_p \omega_p$ following Equation 7.1. Likewise, the speed of a point on the pitch circle of the gear is $v_g = r_g \omega_g$. Because the teeth on the pinion and gear do not slip past or grind against one another, the velocity of points in contact are the same, and $r_g \omega_g = r_p \omega_p$. With the angular velocity of the pinion's shaft being known as the input to the gearset, the speed of the

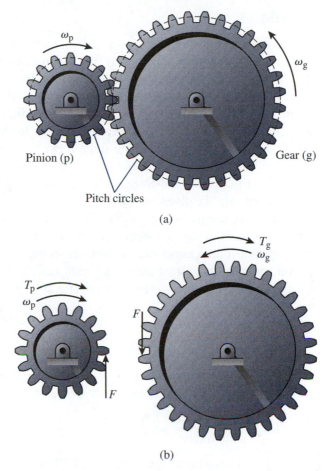

■ **FIGURE 7.8**   (a) A pinion and gear form a gearset, and (b) the pinion and gear are conceptually separated to expose the driving force and the input and output torques.

output shaft in Figure 7.8 is

$$\omega_g = \left(\frac{r_p}{r_g}\right)\omega_p = \left(\frac{N_p}{N_g}\right)\omega_p \tag{7.4}$$

Rather than perform calculations in terms of the pitch radii r_p and r_g, mechanical engineers find it more convenient to work with the numbers of teeth N_p and N_g. While it is a simple matter to count the number of teeth on a gear, it is not so straightforward to measure the pitch radius. Because the diametral pitch (or module) of the pinion and gear are the same, the ratio of radii in Equation 7.4 also equals the ratio of their teeth numbers. The velocity ratio of the gearset is defined as

Velocity ratio
- - - - - - - - - - - ➤

$$VR = \frac{\text{output speed}}{\text{input speed}} = \frac{\omega_g}{\omega_p} = \frac{N_p}{N_g} \tag{7.5}$$

in terms of the output and input speeds.

Torque

We next consider the manner in which torque transfers from the shaft of the pinion to the shaft of the gear. To fix ideas, we imagine that the pinion in Figure 7.8(a) is driven by a motor, and that the shaft of the gear is connected to a crane, pump, or other mechanical load. In the diagrams of Figure 7.8(b), the motor applies torque T_p to the pinion, and torque T_g is transmitted to the shaft of the gear by the load. The driving force F is the physical means by which torque is transmitted between the pinion and gear. The full meshing force between the pinion and gear is the resultant of F and another component that acts along the line of action passing through the two shaft centers. That force component, directed horizontally in Figure 7.8(b), tends to separate the pinion and gear from one another, but it does not contribute to the transfer of torque. As the pinion and gear are imagined to be isolated from one another, F and the separating force act in equal and opposite directions on the teeth in mesh. When the gearset runs at constant speed, the sum of torques applied to the pinion and gear about their centers are each zero, and $T_p = r_p F$ and $T_g = r_g F$. By eliminating the unknown force F, we obtain an expression for the output torque of the gearset:

$$T_g = \left(\frac{r_g}{r_p}\right) T_p = \left(\frac{N_g}{N_p}\right) T_p \tag{7.6}$$

Torque ratio
- - - - - - - - - - - - - - - →

in terms of either the pitch radii or the numbers of teeth. Analogous to Equation 7.5, the torque ratio of the gearset is defined as

$$TR = \frac{\text{output torque}}{\text{input torque}} = \frac{T_g}{T_p} = \frac{N_g}{N_p} \tag{7.7}$$

As you can see from Equations 7.5 and 7.7, the torque ratio for a gearset is exactly the reciprocal of its velocity ratio. If a gearset is designed to increase the speed of its output shaft relative to that of the input shaft ($VR > 1$), then the amount of torque that is transferred will be reduced by an equal factor ($TR < 1$), and vice versa. A gearset exchanges speed for torque, and it is not possible to increase both simultaneously. As a common example of this principle, when an automobile or truck transmission is set in low gear, the rotation speed of the engine crankshaft is reduced by the transmission in order to increase the torque that is applied to the drive wheels.

Power

Following Equation 7.3, the power supplied to the pinion by its motor is $P_p = T_p \omega_p$. On the other hand, the power that is transmitted to the mechanical load by the gear is $P_g = T_g \omega_g$. By combining Equations 7.5 and 7.7, we find that

$$P_g = \left(\frac{N_g}{N_p} T_p\right) \left(\frac{N_p}{N_g} \omega_p\right) = T_p \omega_p = P_p \tag{7.8}$$

which shows that the input and output power levels are the same. The power that is supplied to the gearset is identical to the power removed from it and transferred to the mechanical load. From a practical standpoint, there are frictional losses in any real gearset, but Equation 7.8 is a good approximation for quality gears and bearings and for gearsets where friction is low relative to the overall power level. In short, any reduction of power that occurs from input to output will be associated with frictional losses, and not with the intrinsic nature of changes in speed and torque.

7.4 SIMPLE AND COMPOUND GEARTRAINS

For most combinations of a single pinion and gear as in Figure 7.8, a reasonable limit on the velocity ratio lies in the range of 5 to 10. Larger velocity ratios often become impractical either because of size constraints or because the pinion would need to have too few teeth for smooth engagement with the gear. One might therefore consider building a geartrain that is formed as a serial chain of more that two gears. Such a system is called a *simple geartrain*, and it has the characteristic that each shaft carries a single gear. Figure 7.9 illustrates a simple geartrain having four gears. By convention, the input gear is labeled as gear number 1, and the numbers of teeth and rotational speeds of each gear are represented by N_i and ω_i. The direction that each gear rotates can be determined by inspection, recognizing that for external gearsets, the direction reverses at each mesh point.

Simple geartrain
- - - - - - - - - →

We are interested in determining the geartrain's output velocity, ω_4, relative to its input velocity, ω_1. We view the simple geartrain as a sequence of gearsets that are connected to one another, and we apply Equation 7.5 recursively to each one. Beginning

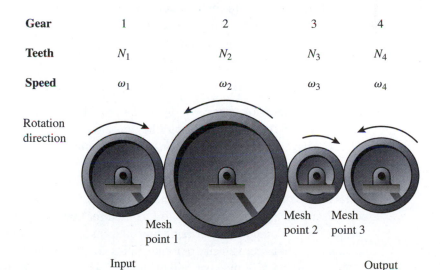

Gear	1	2	3	4
Teeth	N_1	N_2	N_3	N_4
Speed	ω_1	ω_2	ω_3	ω_4

Rotation direction

Mesh point 1

Mesh point 2 Mesh point 3

Input

Output

FIGURE 7.9 A simple geartrain formed as a serial connection of four gears on separate shafts. The gear teeth are omitted for clarity.

at the first mesh point,

$$\omega_2 = \left(\frac{N_1}{N_2}\right)\omega_1 \tag{7.9}$$

with the rotation direction shown in Figure 7.9. At the second mesh point,

$$\omega_3 = \left(\frac{N_2}{N_3}\right)\omega_2 \tag{7.10}$$

By combining these two equations, the intermediate variable ω_2 is eliminated, and

$$\omega_3 = \left(\frac{N_2}{N_3}\right)\left(\frac{N_1}{N_2}\right)\omega_1 = \left(\frac{N_1}{N_3}\right)\omega_1 \tag{7.11}$$

This result shows that the effect of the second gear, and its number of teeth, cancel as far as the third gear is concerned. Proceeding to the final mesh point, we obtain an expression for the velocity of the output gear:

$$\omega_4 = \left(\frac{N_3}{N_4}\right)\omega_3 = \left(\frac{N_3}{N_4}\right)\left(\frac{N_1}{N_3}\right)\omega_1 = \left(\frac{N_1}{N_4}\right)\omega_1 \tag{7.12}$$

For this simple geartrain, the overall velocity ratio between the output and input gears becomes

$$VR = \frac{\omega_4}{\omega_1} = \frac{N_1}{N_4} \tag{7.13}$$

and the sizes of intermediate gears 2 and 3 have no effect on the overall velocity ratio.

Likewise, by applying Equation 7.8 for power conservation with ideal gears, the power that is supplied to the first gear balances the power transferred from the final gear. Since $(VR)(TR) = 1$ for an ideal geartrain, the torque ratio is likewise unaffected by the sizes of gears 2 and 3, and

$$TR = \frac{T_4}{T_1} = \frac{N_4}{N_1} \tag{7.14}$$

Because the intermediate gears in Figure 7.9 provide no speed or torque modifications for the geartrain as a whole, they are sometimes called *idler gears*. While they

Idler gear

have no direct effect on *VR* and *TR*, the idler gears do contribute indirectly to the extent that a designer can insert them to gradually increase or decrease the dimensions of adjacent gears. Additional idler gears also enable the input and output shafts to be separated further from one another, but the power transmission chains and belts discussed in Chapter 3 could also be used for that purpose.

Compound geartrain

Compound geartrains are often used in gearboxes and transmissions when larger velocity or torque ratios are needed or when the system must be compact. In the geartrain of Figure 7.10, the intermediate shaft carries two gears having different numbers of teeth. To determine the overall velocity ratio, ω_4/ω_1, between the input and output of the geartrain, we apply Equation 7.5 to each set of

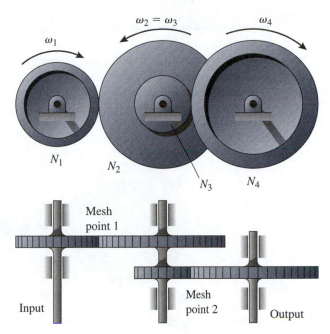

■ **FIGURE 7.10**   A compound geartrain in which the second and third gears are mounted on the same shaft and rotate together. The gear teeth are omitted for clarity.

meshing gears. Beginning with the first pair in mesh,

$$\omega_2 = \left(\frac{N_1}{N_2}\right)\omega_1 \tag{7.15}$$

Because gears 2 and 3 are mounted on the same shaft, $\omega_3 = \omega_2$. Proceeding next to the mesh point between gears 3 and 4, we have

$$\omega_4 = \left(\frac{N_3}{N_4}\right)\omega_3 \tag{7.16}$$

Combining results for the two mesh points, the speed of the output shaft becomes

$$\omega_4 = \left(\frac{N_3}{N_4}\right)\left(\frac{N_1}{N_2}\right)\omega_1 \tag{7.17}$$

and the velocity ratio is

$$VR = \frac{\omega_4}{\omega_1} = \left(\frac{N_1}{N_2}\right)\left(\frac{N_3}{N_4}\right) \tag{7.18}$$

The compound geartrain's overall velocity ratio is the product of the velocity ratios at each mesh point. In contrast to a simple geartrain, the numbers of teeth on the intermediate gears do not cancel. The clockwise or counterclockwise rotation of any gear is determined by inspection and the rule of thumb that with an even number of mesh points, the output shaft rotates in the same direction as the input shaft. Conversely, if the number of mesh points is odd, then the input and output shafts rotate opposite one another.

246 CHAPTER 7 *Motion of Machinery*

EXAMPLE 7.4

The numbers of teeth for gears in the simple geartrain of Figure 7.11 are indicated in the drawing. The input shaft is driven by a motor that supplies 1 kW of power at the operating speed of 250 rpm. (a) Determine the speed and rotation direction of the output shaft. (b) What torque would be transferred from the output shaft to a mechanical load?

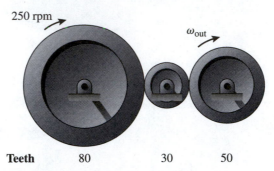

| **Teeth** | 80 | 30 | 50 |

FIGURE 7.11 The simple geartrain of Example 7.4. The gear teeth are omitted for clarity.

SOLUTION

(a) The input shaft rotates clockwise in the figure, and at each mesh point, the direction of rotation reverses. Thus, the 30-tooth gear rotates counterclockwise, and the 50-tooth output gear again rotates clockwise. By applying Equation 7.13, the speed of the output shaft is $(80/50)(250 \text{ rpm}) = 400 \text{ rpm}$.

(b) Neglecting friction, the output power must balance the power that is supplied to the geartrain. Following Equation 7.3, the instantaneous power is the product of torque and rotation speed. We convert the speed of the output shaft to the consistent units of radians per second:

$$\omega_{\text{out}} = (400 \text{ rpm})\left(0.1047 \, \frac{\text{rad/s}}{\text{rpm}}\right) = 41.9 \text{ rad/s}$$

and the output torque becomes

$$T_{\text{out}} = \frac{1000 \text{ W}}{41.9 \text{ rad/s}} = 23.9 \text{ N·m}$$

EXAMPLE 7.5

Figure 7.12 shows the mechanism used in a desktop computer scanner that converts rotation of the shaft of a motor into the side-to-side motion of the scanning head *A*.

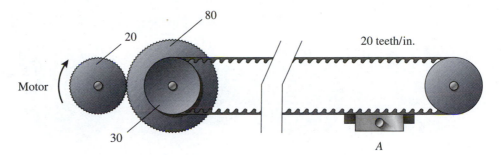

80
20
20 teeth/in.
Motor
30
A

FIGURE 7.12   The geartrain and timing belt mechanism of Example 7.5.

During a portion of a scan operation, the drive motor turns at 180 rpm. The gear attached to the motor shaft has 20 teeth, and the timing belt (Section 3.5) has 20 teeth per inch. The two other gears are sized as indicated. In the units of in./s, calculate the speed at which the scanner head moves.

SOLUTION

The speed of the 80-tooth gear is $(180 \text{ rpm})(20/80) = 45$ rpm. Because the 30- and 80-tooth gears are mounted on the same shaft, the 30-tooth gear also rotates at 45 rpm. With each rotation of that gear, 30 teeth of the timing belt mesh with the gear, and the timing belt advances by the distance

$$\frac{30 \text{ teeth}}{20 \text{ teeth/in.}} = 1.5 \text{ in.}$$

The speed of the timing belt then becomes

$$(1.5 \text{ in./rev})(45 \text{ rpm}) = 67.5 \text{ in./min} = 1.125 \text{ in./s} \qquad \blacksquare$$

7.5   PLANETARY GEARTRAINS

Up to this point, we have examined geartrains in which the shafts are connected to the housing of a gearbox by bearings and the centers of the shafts themselves do not move. The gearsets, simple geartrains, and compound geartrains of the previous sections were each of this type. It is also possible, and desirable, to construct geartrains in which the centers themselves of certain gears move. Such systems are called *planetary geartrains* because their motion is similar in some respects to the orbit of a planet around a star.

Simple and planetary gearsets are contrasted in Figure 7.13. Conceptually, we view the center points of gears in the simple gearset as being connected by a motionless link that is fixed to ground. On the other hand, while the center of the sun gear is stationary in the planetary system of Figure 7.13(b), the center of the planet gear can orbit around the sun gear. The planet gear rotates about its own center, meshes with the sun gear, and orbits as a whole about the

Sun gear

Planet gear

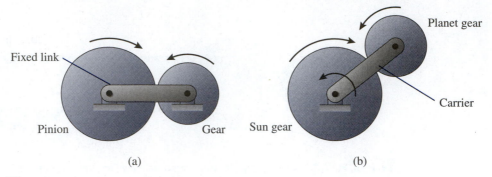

(a)   (b)

■ **FIGURE 7.13**   (a) A simple gearset in which the pinion and gear are connected by a stationary link, and (b) a planetary gearset where the planet and connecting link (or carrier) can rotate about the center of the sun gear.

Carrier
- - - - - - - - - - - - - - - - ➤

Ring gear
- - - - - - - - - - - - - - - - ➤

center of the sun gear. The link that connects the centers of the two gears is called the *carrier*, and it can rotate as well about one of its ends. Planetary geartrains are often used as speed reducers. One application is the geartrain used in the transmission of a light-duty helicopter, as shown in Figure 7.14.

In Figure 7.13(b), it is straightforward enough to connect the shaft of the sun gear to a power source or mechanical load. However, because of the orbital motion of the planet gear, it is not possible to directly connect the planet gear to the shaft of an engine or another machine. To construct a more functional geartrain, the ring gear shown in Figure 7.15 is added to convert the motion of the planet gear into the rotation of the ring gear shaft. The ring gear is an internal gear, whereas the sun and planet gears are external gears. A planetary geartrain therefore has three connection points for input and output, as indicated in Figure 7.16: the shafts of the sun gear, the carrier, and the ring gear. Those three connections can be set to form a geartrain having two input shafts (for instance, the carrier and the sun gear) and one output shaft (the ring gear) or a geartrain having one input and two output shafts. In that

■ **FIGURE 7.14**   A planetary geartrain used in a military helicopter's transmission. The diameter of the ring gear is approximately one foot.
Source: Reprinted with permission of NASA Glenn Research Center.

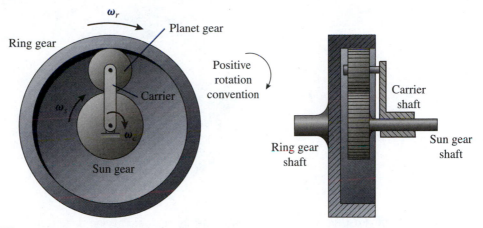

■ FIGURE 7.15   Front and cross-sectional views of a planetary geartrain.

manner, a planetary geartrain can combine power from two sources into one output, or it can split the power from one source into two outputs. In an automobile drivetrain, for instance, power from the driveshaft is split between the two drive wheels by a particular type of planetary geartrain called a *differential*, which we will describe at the end of this section.

One of the geartrain connections can also be fixed to ground (say, the ring gear) so that there is only one input shaft and one output shaft (the carrier and sun gear, in this case). Alternatively, two of the shafts can be coupled together, again reducing the number of connection points from three to two. Versatile configurations such as these are used in automobile automatic transmissions as discussed in Section 8.5. Through the sizing of the sun, planet, and ring gears and through the selection of the input and output connections, engineers can obtain either very small or very large velocity ratios in a geartrain having an otherwise reasonable size.

Balanced geartrain
- - - - - - - - - - - - - - →

Spider
- - - - - - - - - - - - - - →

Planetary geartrains are usually designed with more than one planet gear in order to reduce noise, vibration, and the forces applied to gear teeth. A balanced planetary geartrain is depicted in Figure 7.17. When multiple planet gears are present, the carrier is sometimes called the *spider*, to the extent that it has several (although perhaps not as many as eight) legs, which evenly space the planet gears around the ring gear circumference. In fact, the rolling element bearings discussed in Chapter 3 are analogous to the layout of a balanced planetary geartrain. The motions of the rollers, separator, inner race, and outer race in the tapered roller bearing of Figure 7.18 are similar to those of the planets, carrier, sun gear, and ring gear, respectively, in a planetary geartrain.

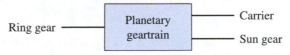

■ FIGURE 7.16   Three input–output connection points to a planetary geartrain.

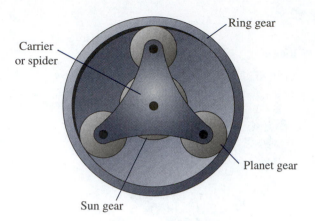

FIGURE 7.17   A balanced planetary geartrain having three planet gears.

FIGURE 7.18   The motions of the rollers, separator, and races in a rolling element bearing are conceptually similar to their counterparts in a planetary geartrain.
Source: Photograph reproduced with permission of The Timken Company.

The rotation of gears and the flow of power through simple and compound geartrains is usually straightforward enough to visualize. Planetary systems are more complicated to the extent that multiple paths exist for power to flow through the geartrain. For instance, if the carrier and sun gear in Figure 7.15 are each driven clockwise, but the carrier speed is greater than that of the sun, the ring gear will also rotate clockwise. However, as the speed of the sun gear would be gradually increased, the ring gear would slow down, stop, and then reverse direction to rotate counterclockwise.

Our intuition is generally not sufficient to determine the direction that each gear and shaft in a planetary geartrain will rotate. For that reason, we instead apply a design equation that relates the rotational velocities of the sun gear (denoted ω_s), carrier (ω_c), and ring gear (ω_r) as follows:

$$(2+n)\omega_r + n\omega_s - 2(1+n)\omega_c = 0 \tag{7.19}$$

The direction that each component rotates is determined from the positive or negative sign of the corresponding angular velocity term. As shown in Figure 7.15, we apply the

Sign convention
- - - - - - - - - - - - - - - ➤

convention that rotations of the gears and carrier are positive when directed clockwise, and negative when counterclockwise. By consistently applying the sign convention, we rely on the outcome of the calculation to indicate the directions that the components rotate. With the numbers of teeth on the sun and planet gears denoted by N_s and N_p, the form factor n in Equation 7.19 is

Form factor
- - - - - - - - - - - - - - - ➤

defined as

$$n = \frac{N_s}{N_p} \tag{7.20}$$

This parameter is not the geartrain velocity or torque ratio, nor is it the number of teeth on any gear. It is simply a size ratio that enters into the speed calculations for planetary geartrains and makes them more convenient. Equations 7.19 and 7.20 are derived in Appendix C, and in the following two examples, we apply them to calculate shaft speeds and rotation directions.

EXAMPLE 7.6

Figure 7.19 shows the cross section of a planetary geartrain having two inputs (sun gear and carrier) and one output (ring gear). When viewed from the right-hand side in Figure 7.19, the hollow carrier shaft is driven at 3600 rpm clockwise and the shaft for the sun gear turns at 2400 rpm counterclockwise. (a) Determine the speed and direction of rotation of the ring gear. (b) The sun gear is driven by an electric motor that is capable of reversing its rotation direction. Repeat the calculation for the case in which the sun gear is instead driven clockwise at 2400 rpm. (c) For what speed and direction of the sun gear would the ring gear not rotate?

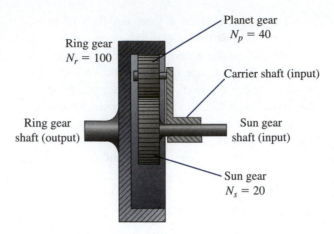

FIGURE 7.19   Planetary geartrain with two input shafts and one output shaft in Example 7.6.

SOLUTION

(a) We solve this problem by applying Equations 7.19 and 7.20. With the sign convention noted in Figure 7.15, we have $\omega_c = 3600$ rpm and $\omega_s = -2400$ rpm. Because this calculation involves only rotational speeds and not the velocity of a point as in Equation 7.1, it is acceptable to retain the units rpm rather than convert angular velocity to radians per unit time. Based on the numbers of teeth listed in the figure, the form factor is $n = 20/40 = 0.5$. The rotation speed of the ring gear becomes

$$\omega_r = \frac{2(1+0.5)}{2+0.5}(3600 \text{ rpm}) - \frac{0.5}{2+0.5}(-2400 \text{ rpm})$$

$$= 4320 \text{ rpm} + 480 \text{ rpm}$$

$$= 4800 \text{ rpm}$$

Because $\omega_r > 0$, the ring gear turns clockwise, the same direction as the carrier.

(b) If the direction of the sun gear is reversed, $\omega_s = 2400$ rpm, and the speed of the ring gear now becomes

$$\omega_r = 4320 \text{ rpm} - 480 \text{ rpm} = 3840 \text{ rpm}$$

The ring gear rotates in the same direction as in (a), but at a 20% lower speed.

(c) The ring gear does not rotate when $\omega_r = 0$. At that condition in Equation 7.19, the speeds of the carrier and sun gear are related by

$$n\omega_s - 2(1+n)\omega_c = 0$$

With $\omega_c = 3600$ rpm, the speed of the sun gear at this condition is $\omega_s = 21,600$ rpm. The result is positive, so the motor drives the sun gear clockwise. ■

EXAMPLE 7.7

For the geartrain of Example 7.6(b), the sun gear and carrier are driven by engines producing 5 hp and 2 hp, respectively. Determine the torque that is applied to the shaft of the ring gear.

SOLUTION

The output shaft transfers a total of 7 hp to balance the power that is supplied to the geartrain. Expressing the speed of the ring gear in consistent units,

$$\omega_r = (3840 \text{ rpm}) \left(0.1047 \frac{\text{rad/s}}{\text{rpm}} \right) = 402 \text{ rad/s}$$

and the output torque is calculated by applying Equation 7.3:

$$T = \frac{(7 \text{ hp})(550 \text{ (ft·lb/s)/hp})}{402 \text{ rad/s}} = 9.57 \text{ ft·lb}$$ ■

Differential
- - - - - - - - - - - - - - →

An automobile differential is a special type of planetary geartrain. The layout of the drivetrain in a rear-wheel-drive vehicle is shown in Figure 7.20.

With the engine located at the front, the crankshaft feeds into the transmission, which in turn is connected to a driveshaft that extends down the length of the vehicle. The transmission adjusts the velocity ratio between the rotation speeds of the engine crankshaft and the driveshaft. The differential is a geartrain that transfers torque from the (input) driveshaft and splits it between the left and right rear drive wheels (the outputs) of the vehicle. Take a look underneath a rear-wheel-drive automobile, and see if you can identify the transmission, driveshaft, and differential.

The differential enables the wheels that are powered by the engine to rotate at different speeds when the vehicle turns a corner. The drive wheel on the outside of a curve must roll through a greater distance than the wheel on the inside. If the drive wheels were rigidly connected to one another and constrained to always rotate at the same speed, they would need to slide and skid during the turn. In order to solve that problem, mechanical engineers developed the differential to be placed between the driveshaft and the axles to the rear wheels. The differential enables the wheels to rotate at different speeds as the vehicle follows curves along a road, even while being connected to and powered by the driveshaft. Figure 7.21 illustrates the construction of an automobile differential. The carrier is a structural shell that is attached to an external ring gear which, in turn, meshes with a pinion on the driveshaft. As the gear and housing of the carrier rotate with the driveshaft, the two planet gears orbit in and out of the plane of the figure. Because of the bevel-type gear construction, the ring and sun gears of an equivalent planetary geartrain would have the same pitch radius. For that reason, the ring and sun gears in a differential are interchangeable, and both are conventionally called sun gears.

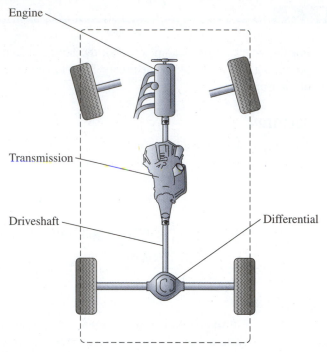

■ FIGURE 7.20   Arrangement of the drivetrain in a rear-wheel-drive automobile.

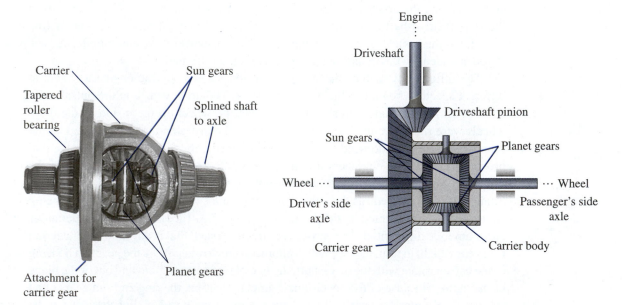

■ FIGURE 7.21   A differential for a small passenger automobile, which can be viewed as a planetary geartrain that is constructed from bevel-type gears.

7.6 ENGINE AND COMPRESSOR MECHANISMS

As we introduced during the discussion of internal combustion engines in Section 6.6, the piston, connecting rod, and crankshaft convert the back-and-forth motion of the piston into rotation of the crankshaft. This mechanism is also the basis for reciprocating pumps and compressors, in which the crankshaft is instead driven by a power source and the piston performs work as it moves a liquid or compresses a gas. In either case, this mechanism changes its shape as sketched in Figure 7.22, and it smoothly converts the motion of the piston along the cylinder into rotation of the crankshaft. In the process of designing mechanisms of this type, mechanical engineers simulate the motions of the piston, connecting rod, and crankshaft, and they calculate the forces that act on those components. Of particular interest are the velocity and acceleration of the piston when the engine or compressor operates at a steady speed. Practicing engineers encounter many different types of mechanisms in their work,

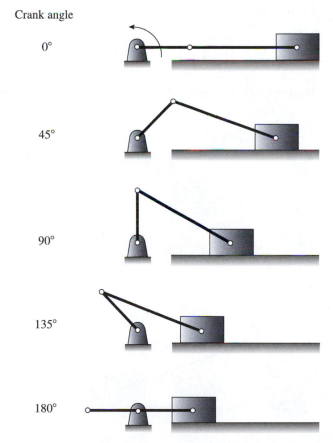

FIGURE 7.22 Visualizing motion of the piston, connecting rod, and crankshaft mechanism as the crank rotates through 180°.

and in this section we explore one example in order to introduce you to the issues at hand.

Figure 7.23 labels the geometry of the mechanism. The rotation angle θ and speed ω of the crankshaft are taken as known quantities. Our objective is to calculate the position x and velocity v of the piston along the cylinder as the crankshaft rotates. The motion of the piston can be fully determined by applying trigonometric identities to the triangle ABC shown in Figure 7.23(c). The horizontal position of the piston is

$$x = r \cos \theta + L \cos \phi \tag{7.21}$$

where r and L are the lengths of the crank and connecting rod, respectively. The connecting rod angle ϕ (the lowercase Greek character phi) is not known explicitly at this point, but we can calculate it in terms of the crankshaft angle. The height h in Figure 7.23(c) is given by both $r \sin \theta$ and $L \sin \phi$, so the connecting rod angle is related to the crankshaft angle by $\sin \phi = (r/L) \sin \theta$. Using the identity $\cos^2 \phi + \sin^2 \phi = 1$, the term for the connecting rod angle in Equation 7.21 becomes

$$\cos \phi = \sqrt{1 - \left(\frac{r}{L}\right)^2 \sin^2 \theta} \tag{7.22}$$

By substituting Equation 7.22 into Equation 7.21, we eliminate ϕ and obtain an equation

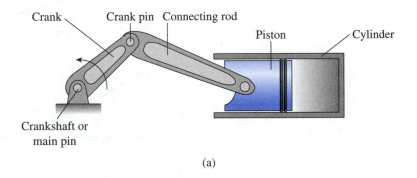

(a)

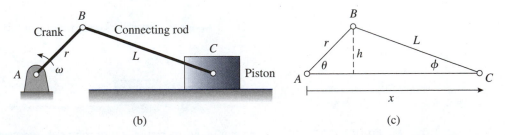

(b)   (c)

■ FIGURE 7.23   The changing shape of the piston, connecting rod, and crankshaft mechanism.

for the position of the piston as a function of the crankshaft angle:

$$x = r \cos \theta + L\sqrt{1 - \left(\frac{r}{L}\right)^2 \sin^2 \theta} \tag{7.23}$$

For practical designs, r must be less than L or else the piston would contact the crankshaft, and so the subtraction operation within the radical does not pose a problem.

The velocity of the piston is defined as the time rate of change in its position. To obtain an expression for the velocity, we use the principles of calculus and differentiate Equation 7.23 with respect to time. The piston velocity then becomes $v = dx/dt$.

Chain rule
- - - - - - - - - - - - →

Because x is a function of the crankshaft angle but θ itself changes with time, we apply the chain rule for differentiation. For the function $f(\theta)$ with θ changing in time, the derivative of f is

$$\frac{df}{dt} = \left(\frac{df}{d\theta}\right)\left(\frac{d\theta}{dt}\right) \tag{7.24}$$

In our case, $d\theta/dt = \omega$ is the crankshaft rotation speed. The velocity of the piston can therefore be found by differentiating Equation 7.23 with respect to θ and multiplying the result by ω. By applying the differentiation properties reviewed in Appendix B, our result becomes

$$v = -r\omega \, \sin \theta \left(1 + \frac{(r/L) \cos \theta}{\sqrt{1 - (r/L)^2 \sin^2 \theta}}\right) \tag{7.25}$$

where the units of ω are radians per unit time. When the lengths of the crank and connecting rod, the crankshaft speed, and the angle of the crankshaft are known, Equation 7.25 can be used to calculate the velocity of the piston. Figure 7.24 illustrates how the

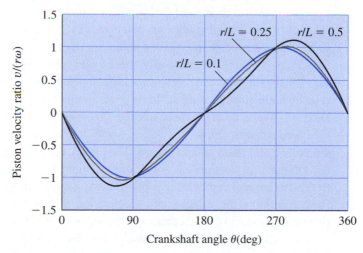

FIGURE 7.24 Velocity ratio of the piston (v) and the crank ($r\omega$) for ratios $r/L = 0.1$, 0.25, and 0.5.

velocity varies during one revolution of the crankshaft for several values of the crank ratio r/L. As r/L grows, the shape of the velocity curve of the piston becomes slightly distorted relative to a sine wave.

EXAMPLE 7.8

What is the average velocity of the piston during its stroke from the bottom of the cylinder to the top?

SOLUTION

During one stroke, the piston moves through a total distance of $2r$ while the crankshaft rotates through 180° (or π rad). The time required for one-half rotation of the crankshaft to occur is π/ω. The average velocity of the piston during one stroke is

$$v_{\text{avg}} = \frac{2r}{\pi/\omega} = 0.636r\omega \tag{7.26}$$

Referring to Figure 7.23, $r\omega$ is the (constant) velocity of point B on the crankshaft. During one stroke, the piston therefore moves on average at about 64% of the crank pin's speed.   ∎

SUMMARY

In the previous sections, we examined the motion of machinery as a fifth element of mechanical engineering. Important quantities introduced in this chapter, common symbols representing them, and their units are summarized in Table 7.2. Gearsets, simple geartrains, compound geartrains, and planetary geartrains are used for the purposes of transmitting power, changing the rotation speed of a shaft, and modifying the torque that is applied to a shaft. More broadly, mechanisms are combinations of links, shafts, bearings, springs, cams, lead screws, and other building-block components that convert one type of motion into another. A "library" comprising recipes for thousands of other mechanisms is available in the mechanical engineering literature for the design of automatic feed mechanisms, conveyor systems, safety latches, ratchets, self-deploying aerospace structures, and many other systems. The piston, connecting rod, and crankshaft assembly analyzed in Section 7.6 is just one example of a practical mechanism that arises in the mechanical engineering profession. At a higher level, the application of calculus for determining the velocity and acceleration of a piston should give you a perspective on the impact of mathematics in the engineering curriculum. Among other topics, the motion of machinery encompasses mechanical components, forces and torques, and energy and power, and it also serves as a transition to the capstone subject of mechanical design in the following chapter.

TABLE 7.2 Quantities, Symbols, and Units That Arise in the Analysis of Motion and Machinery

Quantity	Conventional Symbols	Conventional Units	
		USCS	SI
Velocity	v	in./s, ft/s	mm/s, m/s
Angle	θ, ϕ	deg, rad	
Angular velocity	ω	rpm, rps, deg/s, rad/s	
Torque	T	in·lb, ft·lb	N·m
Work	ΔW	ft·lb	J
Instantaneous power	P	hp, ft·lb/s	W
Velocity ratio	VR	—	—
Torque ratio	TR	—	—
Form factor	n	—	—

SELF-STUDY AND REVIEW

1. List several different units that are used for angular velocity.
2. In what types of calculations must one use the unit of rad/s for angular velocity?
3. What is the difference between average and instantaneous power?
4. How do simple and a compound geartrains differ?
5. How are the velocity and torque ratios of a geartrain defined?
6. What relationship exists between the velocity and torque ratios for an ideal geartrain?
7. Sketch a planetary geartrain, label its main components, and explain how it operates.
8. What is a balanced planetary geartrain?
9. Describe the main components of an automobile drivetrain.
10. What function does the differential in an automobile serve?
11. Sketch the piston, connecting rod, and crankshaft mechanism in an internal combustion engine or compressor.
12. How are differentiation and the chain rule of calculus used when engineers calculate the velocity and acceleration of mechanisms?

PROBLEMS

1. An automobile travels at 30 mph, which is also the speed of the center C of the tire. If the outer radius of the tire is 15 in., determine the rotational velocity of the tire in the units of rad/s, deg/s, rps, and rpm.

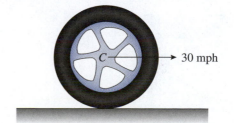

2. The surface of a disk in a computer hard drive is shown in the accompanying figure. The disk spins at 7200 rpm. At the radius of 30 mm, a stream of data is magnetically written on the disk. The spacing between data bits is 25 μm. Determine the number of bits per second that pass by the read/write head.

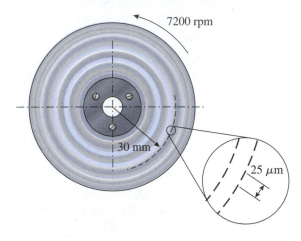

3. In the top view of an industrial robot shown in the figure, the lengths of the two links are $AB = 22$ in. and $BC = 18$ in. For a particular maneuver, the angle between the two links is held constant at 40° as the robot's

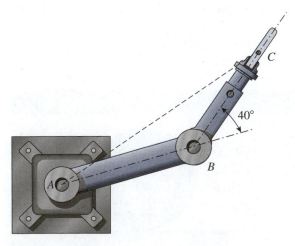

arm rotates about base A at 300 deg/s. Calculate the velocity of the center C of the gripper.

4. A gasoline-powered engine operates at the maximum rating of 15 hp, and it drives a water pump at a construction site. If the engine speed is 450 rpm, determine the torque T that is transmitted from the output shaft of the engine to the pump, neglecting any frictional losses. Express your answer in the units of ft·lb and N·m.

5. A small automobile engine produces 260 N·m of torque at 2100 rpm. Determine the power output of the engine in the units of kW and hp.

6. The torque produced by a 2.5-liter automobile engine, operating at full throttle, was measured over a range of engine speeds. By using the accompanying graph of torque as a function of speed, prepare a second graph to show how the power output of the engine (in the units of hp) changes with speed (in rpm).

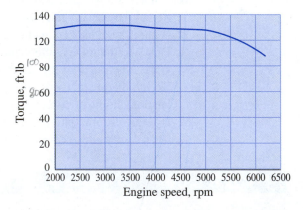

7. The helical gears in the simple geartrain shown in the figure on the next page have teeth numbers as labeled. The central gear rotates at 125 rpm and drives the two output shafts. Determine the speeds and rotation directions of each shaft in the geartrain.

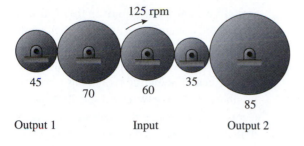

125 rpm

45 70 60 35 85

Output 1   Input   Output 2

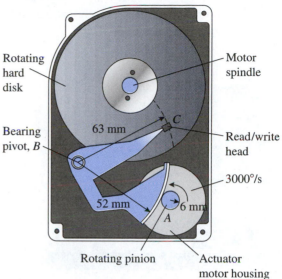

Rotating hard disk

Motor spindle

Bearing pivot, B

63 mm

C

Read/write head

3000°/s

52 mm 6 mm

A

Rotating pinion

Actuator motor housing

8. The spur gears in the compound geartrain shown in the figure have teeth numbers as labeled. (a) Determine the speed and rotation direction of the output shaft. (b) If the geartrain transfers 4 hp, calculate the torque that is applied to the input and output shafts.

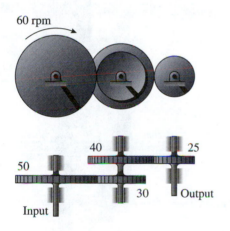

60 rpm

40 25

50

30 Output

Input

9. The disk in a computer magnetic hard drive shown in the figure spins about its center while the recording head C reads and writes data. The head is positioned above a specific data track by the arm, which pivots about its bearing at B. As the actuator motor A turns through a limited angle of rotation, its 6-mm radius pinion rotates along the arc-shaped segment, which has a radius of 52 mm about B. If the actuator motor turns at 3000°/s over a small range of motion during a track-to-track seek operation, calculate the speed of the read/write head.

10. For the compound geartrain shown in the figure, develop an expression for the velocity and torque ratios in terms of the numbers of teeth labeled on the diagram.

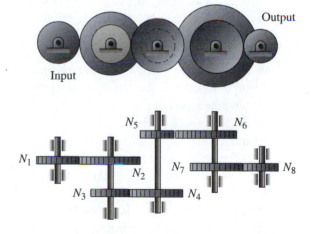

Output

Input

N_5 N_6

N_1 N_2 N_7 N_8

N_3 N_4

11. The motorized rack and pinion system within a milling machine is shown in the figure on the next page. The pinion has 50 teeth and a pitch radius of 0.75 in. Over a certain range of its motion, the motor turns at 800 rpm. (a) Determine the horizontal velocity of the rack. (b) If a peak torque of 10 ft·lb is supplied to

the motor, determine the force that is in turn supplied to the rack.

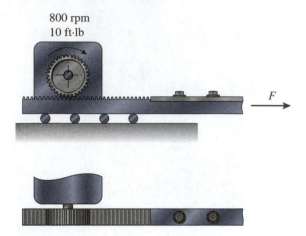

800 rpm
10 ft·lb

F

12. (a) By directly examining sprockets and counting their numbers of teeth, determine the velocity ratio between the (input) pedal crank and the (output) rear wheel for your, a friend's, or a family member's multiple-speed bicycle. Make a table to show how the velocity ratio changes depending on which sprocket is selected in the front and back of the chain drive. Tabulate the velocity ratio for each speed setting. (b) For a bicycle speed of 15 mph, determine the speed of the sprockets, and the speed of the chain, for one of the chain drive speed settings.

13. Estimate the velocity ratio between the (input) engine and the (output) drive wheels for your, a friend's, or a family member's automobile in several different transmission settings. You will need to know the engine speed (as read on the tachometer), the vehicle speed (on the speedometer), and the outer diameter of the wheels. Show in your calculations how the units of the various terms are converted in order to obtain a dimensionally consistent result.

14. In the two-speed gearbox shown in the figure, the lower shaft is splined and can slide hori-

zontally as an operator moves the shift fork. Design the transmission and choose the number of teeth on each gear in order to produce velocity ratios of 0.8 (in first gear) and 1.2 (in second gear). Select gears having only even numbers of teeth between 40 and 80. Note that a constraint exists on the sizes of the gears so that the locations of the shaft centers are the same in the first and second gears.

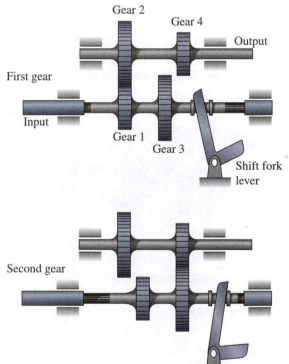

Gear 2
Gear 4
Output
First gear
Input
Gear 1
Gear 3
Shift fork lever
Second gear

15. Explain why the number of teeth on the ring gear (N_r) in a planetary geartrain is related to the sizes of the sun and planet gears by $N_r = N_s + 2N_p$.

16. The geartrain of Figure 7.15 has $N_s = 48$ and $N_p = 30$, and it uses the carrier and ring gear as inputs. The sun gear is the output. When viewed from the right-hand side, the hollow carrier shaft is driven at 1200 rpm clockwise, and the shaft for the ring gear is driven at 1000 rpm counterclockwise. (a) Determine the speed and direction of rotation of the sun gear. (b) Repeat the calculation for the case in which the carrier is instead driven clockwise at 2400 rpm.

17. The outer race of the straight roller bearing in the accompanying figure is held by a pillow block mount. The inner race supports a shaft that rotates at 1800 rpm. The radii of the inner and outer races are $R_i = 0.625$ in. and $R_o = 0.875$ in. In the units of ms, how long does it take for "roller 1" to orbit the shaft and return to the topmost position in the bearing? Does the roller orbit in a clockwise or counterclockwise direction?

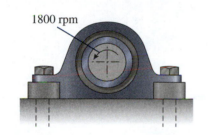

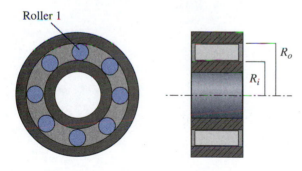

18. (a) The shaft of the sun gear in the planetary geartrain on the accompanying figure is held stationary by a brake. Determine the relationship between the rotational speeds of the shafts for the ring gear and carrier. Do those shafts rotate in the same or opposite direction? (b) Repeat the exercise for the case in which the ring gear shaft is instead held sta-

tionary. (c) Repeat the exercise for the case in which the carrier shaft is instead held stationary.

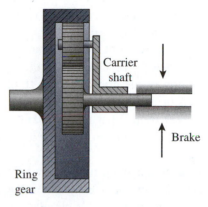

19. In one of the planetary geartrain configurations that arises in the design of automobile automatic transmissions, the shafts for the sun gear and carrier are connected to one another and turn at the same speed, ω_o, as shown in the accompanying figure. Determine the speed of the ring gear shaft for this configuration.

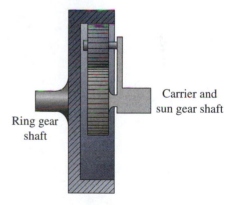

20. For the crank and connecting rod lengths of $r = 1$ in. and $L = 7$ in. and the engine speed of 2000 rpm, prepare a graph of the velocity of the piston (in the units of ft/s) as a function of the crankshaft angle (in degrees). What is the maximum velocity, and at what crankshaft angle does it occur?

21. Differentiate Equation 7.25 with respect to time and obtain an expression for the acceleration of the piston as a function of the crank angle. Use the chain rule for differentiation, and specify that the crankshaft rotation speed is constant $(d\omega/dt = 0)$.

22. Applying the result from Problem 21, prepare a graph of the acceleration of the piston as a function of the crank angle. Use the values $r = 1$ in. and $L = 7$ in. and the engine speed of 2000 rpm. What is the maximum acceleration of the piston, expressed in the units of g's?

23. The accompanying figure shows the preliminary design for a cam and valve mechanism. The cam itself is circular with radius $r = 15$ mm, but its center C is offset from the shaft's center O by $e = 4$ mm. The shaft turns counterclockwise at the constant speed 1000 rpm. (a) Determine an expression for the vertical position x of the valve as a func-

tion of the cam's rotation angle θ. (b) Differentiate your result to obtain expressions for the velocity and acceleration. (c) What are the maximum velocity and acceleration of the valve, and at what angles θ of the cam do they occur?

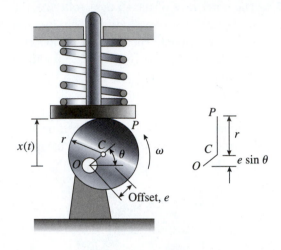

8 Mechanical Design

8.1 OVERVIEW

While discussing the differences among engineers, scientists, and mathematicians in Chapter 1, we saw that the word *engineering* is related to both *ingenious* and *devise*. Creative design lies at the center of the mechanical engineering profession, and an engineer's ultimate goal is to produce new hardware that solves one of society's technical problems. Beginning either from a blank sheet of paper or from existing hardware that is being modified, the product development process often forms the focus of an engineer's activities. In keeping with their profession's title, many engineers truly are ingenious, and they possess the vision and skill to make such lasting contributions as those described in the top ten list of Section 1.3.

Formal education in engineering is not a prerequisite to having a good idea for a new or improved product. Your interest in studying mechanical engineering, in fact, may have been sparked by your own ideas for building hardware. The elements of mechanical engineering that we have examined up to this point—machine components and tools, forces in structures and fluids, materials and stresses, thermal and energy systems, and the motion of machinery—are intended to have set a foundation that will enable you to approach mechanical design in a more effective and systematic manner. In that respect, the approach taken in this textbook is a condensed analog of the traditional engineering curriculum: Approximation, mathematics, and science are applied to design problems in order to increase performance and reduce trial and error. By applying the resources of Chapters 2–7, you can select certain machine components and perform back-of-the-envelope calculations to guide design decisions. Such analyses are not made for their sake alone; rather, they enable you to design better and faster.

Effective mechanical design is a broad area, and the creative and technical processes behind it cannot be set forth fully in one chapter—or even one textbook for that matter. Indeed, with this material as a starting point, you should continue to develop hands-on experience and design skills throughout your entire professional career. Even the most seasoned engineer grapples with the procedure for transforming an idea into manufactured hardware that can be sold at a reasonable cost.

After first discussing the hierarchy of steps that engineers take when they transform a new idea into reality, we explore the subject of mechanical design through three case studies in the fields of conceptual design, computer-aided design, and detailed machine design. We will also discuss mechanical design from a business perspective and describe how patents protect newly developed technology. After completing this chapter, you should be able to:

- Outline the major steps and iteration points in the high-level mechanical design procedure.
- Give an example of the processes for brainstorming and for identifying the advantages and disadvantages of various design options.
- Understand the role played by computer-aided engineering tools in mechanical design, and describe how such tools can be seamlessly integrated with one another.
- By using a sketch as a guide, describe the operation of an automobile automatic transmission, a complex machine design that incorporates mechanical, electronic, computer, and hydraulic components.
- Explain what patents are, and discuss their importance to engineering's business environment.

8.2 HIGH-LEVEL DESIGN PROCEDURE

In this section, we outline the steps that engineers take when they develop new products and hardware. From the broadest viewpoint, design is defined as the systematic process for devising a mechanical system to meet one of society's technical needs. The specific motivation could lie in the areas of transportation, communication, or security, for instance. The prospective product is expected to solve a particular problem so well, or offer such a new capability, that others will pay for it. Early on, a company's marketing department will collaborate with engineers and managers to identify, in a general sense, new opportunities for products. Together, they define the new product's concept by drawing upon feedback from potential customers and from users of related products. Designers will subsequently develop those concepts, work out the details, and bring the functioning hardware to realization. Many approximations, trade-offs, and choices are made along the way, and mechanical engineers are mindful that the level of precision that is needed will naturally and gradually grow as the design matures. For instance, it does not make sense for an engineer to resolve specific details (Should a grade 1020 or 1045 steel alloy be used? Are ball or roller bearings most appropriate? What must be the viscosity of the oil?) until the design's overall concept has taken firm shape. After all, at an early stage of the design cycle, the specifications for the product's size, weight, power, or performance could still change. Design engineers are comfortable with such ambiguity, and they are able to develop products even in the presence of requirements and constraints that can change.

Creativity
- - - - - - - - - - - - - →

Simplicity
- - - - - - - - - - - - - →

Iteration
- - - - - - - - - - - - - →

The formal procedure by which a marketing concept evolves into manufactured hardware is based upon many principles and attributes. Most engineers would probably agree that creativity, simplicity, and iteration are key factors in any successful endeavor. Innovation begins with a good idea, but it also implies starting from a blank sheet of paper. Nevertheless, engineers must still take the first, perhaps uncertain, step for transforming that formative idea into concrete reality. Early design decisions are made by drawing upon a variety of sources: personal experience, knowledge of mathematics and science, laboratory and field testing, and trial and error guided by good judgment. Generally speaking, simpler design concepts are better than complex ones, and the adage "keep it simple, stupid" has a well-deserved reputation among engineers for guiding decisions. Iteration is also important for improving a design and for refining hardware that works into hardware that works well. The first idea that you have, just like the first prototype that you construct, will probably not be the best ones that can be realized. With the gradual improvement of each iteration, however, the design will perform better, more efficiently, and more elegantly.

From a macroscopic perspective, the mechanical design procedure can be broken down into four major steps, which are outlined with greater detail in Figure 8.1.

1. **Define and research objectives.** Initially, a designer describes the new product's requirements in terms of its function, weight, strength, cost, safety, reliability, and so forth. At this first stage, constraints that the design must satisfy are also established. Those constraints might be of a technical nature—say, a restriction on size or power consumption. Alternatively, the constraints could be related to business or marketing concerns, such as the product's appearance, cost, or ease of use. When faced with a new technical challenge, engineers will conduct research and gather background information that is expected to be useful when concepts and details are later evaluated. Engineers read patents that have been issued for related technologies, consult with vendors of components or subsystems that might be used in the product, attend expositions and trade shows, and meet with potential customers to better understand the application. Early in the design process, engineers define the problem, set the objective, and gather pertinent information for the foundation of a good design.

2. **Generate concepts.** In this stage, designers generally work in teams with the goal of devising a wide range of potential solutions to the problem at hand. This creative effort involves conceiving new ideas and combining previous ones to be greater than the sum of their parts. Hardware solutions are conceptualized and composed, and both good and not-so-good ideas are tossed about. Results from the brainstorming sessions are systematically recorded, the advantages and disadvantages of various solutions are identified, and trade-offs among the differing approaches are made. To document the suite of ideas that emerges from this synthesis stage, engineers sketch concepts, make notes, and prepare lists of "pros and cons" in their design notebooks. No particular idea is evaluated in depth, nor is any idea viewed with too critical an eye. Instead, you should focus on cataloging multiple approaches and devising a wide range of design concepts,

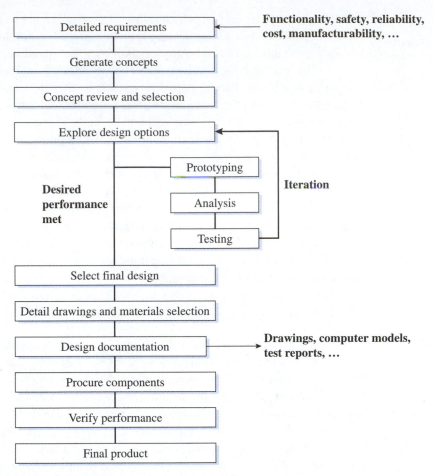

FIGURE 8.1 Flowchart of the prototypical mechanical design procedure.

not necessarily all conventional ones. Even though a particular solution might not seem feasible at this early stage, should the product's requirements or constraints change in the future (which is likely), the idea might in fact resurface as a leading contender.

3. **Narrow down the options.** The design team further evaluates the concepts with a view toward reducing them to a promising few. For instance, engineers make preliminary calculations to compare strength, safety, cost, and reliability, and they will begin to discard the less feasible concepts. Sample hardware could also be produced at this stage. Just as a picture is worth a thousand words, a physical prototype is often useful for engineers to visualize complex machine components and to explain their assembly to others. The prototypes can also be tested so that trade-off decisions are made based on the results of both measurements and analyses. One method for producing such components is called *rapid prototyping*, and its key capability is that complex, three-dimensional components can be fabricated directly from a

Rapid prototyping

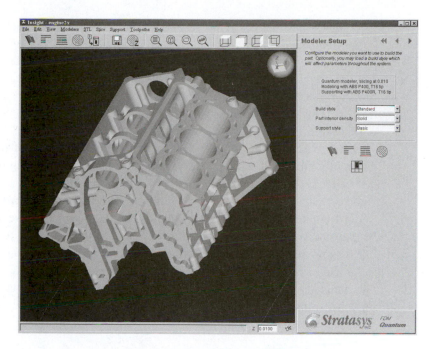

(a)

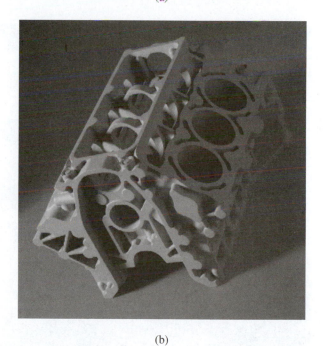

(b)

■ **FIGURE 8.2**   (a) Screen capture of a computer-aided design drawing for an engine block. (b) Physical model produced through rapid prototyping.

Source: Reprinted with permission of Stratasys, Incorporated.

computer-generated drawing, often in a matter of hours. One such technology is called *fused deposition modeling*, and it enables durable and fully functional prototypes to be fabricated from plastics and polycarbonates. As an example, Figure 8.2 depicts a computer-aided design drawing of an engine block and a physical prototype developed with the system shown in Figure 8.3.

4. **Develop a detailed design.** To reach this point of the high-level procedure, the design team will have brainstormed, tested, analyzed, and converged its way to what it perceives as the best concept. The implementation of the design, construction of a final prototype, and development of the manufacturing process each remain. Detailed technical issues are solved by applying mathematical, scientific, laboratory, and computer-aided engineering tools. Completed drawings and parts lists are prepared. The designers conduct engineering analysis and experiments to verify performance over a range of operating conditions. If necessary, changes to shape, dimensions, materials, and components will be made until all requirements and constraints are met. The design is documented through engineering drawings and written reports so that others are able to understand the reasons behind each of the many decisions that the designers made. Such documentation is also useful for future design teams to learn from and build upon the present team's experiences.

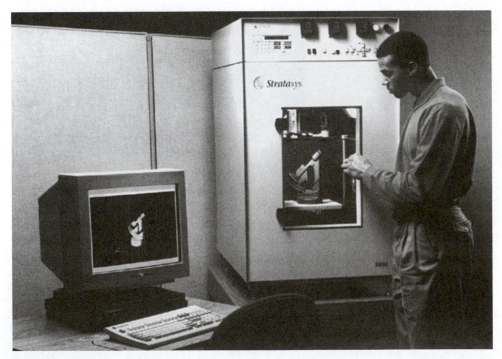

■ **FIGURE 8.3**　A fused deposition system for rapid prototyping.
Source: Reprinted with permission of Stratasys, Incorporated.

At the most fundamental level, the final design must meet all of its requirements and constraints. You might think that an engineer's tasks are completed once the working prototype has been delivered or after the finishing touches have been applied to the drawings. However, mechanical engineers today work in a broader environment, and their hardware is viewed with a critical eye beyond the criterion of whether or not it functions as intended. For a product to be successful, it must also be safe to use, reliable, environmentally sound in its use and disposal, and affordable to manufacture. After all, if the product is technically superb but it requires expensive materials and manufacturing operations, customers may avoid the product and select one that is more balanced in cost and performance. In the end, engineering is a business venture that must meet the needs of its customers.

8.3 CASE STUDY IN CONCEPTUAL DESIGN: MOUSETRAP-POWERED VEHICLES

As applied to engineering, the adage "well begun is half done" reinforces the importance of creative brainstorming in the conceptual phase of the high-level design procedure. By way of a case study in this section, we trace the progress of a hypothetical team of engineering students as they generate concepts for designing a mousetrap-powered vehicle. Small toy vehicles that are powered by household mousetraps are sometimes the subject of design and construction projects in engineering and science courses. Readily visualized and built, these vehicles are a useful means for experiencing the conceptual design process and for gaining an appreciation for the trade-offs that must be made for a design to satisfy all of its constraints.

In our illustrative case study, teams of engineering students are challenged to build small vehicles that travel 10 meters as quickly as possible but that are powered by only the potential energy stored in a mousetrap spring. Each vehicle will be designed, built, refined, and operated by a team of three students. At the conclusion of the project, the teams will compete against one another in head-to-head races during a tournament, so the final products must be both durable and reusable. In addition to the overall objective of producing a fast vehicle, several other specifications must be met:

1. The mass of the vehicle cannot exceed 500 g.
2. The vehicle must fit completely within a 0.1-m^3 box at the start of each race.
3. Each vehicle will race in a lane that is 10 m long but only 1 m wide. If any part of the vehicle passes outside the lane during a race, the team will be disqualified from the tournament.
4. The vehicle must remain in contact with the surface of the lane during the entire race.
5. The vehicle must be powered only by a standard household mousetrap. Energy that is incidentally stored by other elastic elements or obtained from a change in elevation of the vehicle's center of mass must be negligible.
6. Tape cannot be used as a fastener in the vehicle's construction.

Each of these specifications constrains, in different ways, the hardware that the teams will ultimately produce. If any single requirement is not met, the entire design will be inadequate, regardless of how well the vehicle might perform relative to the other requirements. For instance, because the racing lane is 10 times longer than it is wide, the vehicle must not only be fast but must also travel in a reasonably straight line. If a particular vehicle sometimes veers outside the lane, then it could be defeated by a slower vehicle (even a much slower one) in a head-to-head race. The design teams recognize that the vehicles should not be optimized with respect to only one specification, but rather should be balanced to meet all requirements.

Design notebook
- - - - - - - - - - - - - - →

We next follow the thought process of a hypothetical team as it begins to brainstorm and identify multiple design concepts. The students document their ideas in a bound notebook, and they use written comments and hand drawings to describe each concept. Subsequently, the team will record progress as prototypes are constructed and the outcome of their testing and iteration efforts. In short, the notebook serves as a log to chronicle the team's entire design experience. In industrial research and development settings, such notebooks are often dated, signed, and even witnessed in order to formally document a product's development. With an eye toward your own professional career, you should also begin the practice of systematically recording, revising, and developing your original ideas.

First Concept: String and Lever Arm

An idea that emerges from the team's first brainstorming session is based on using the mousetrap's snap arm to pull and unwrap string from a drive axle. Together, the team members sketch the concept shown in Figure 8.4. As the trap snaps shut, string is unwound from a spool that is attached to the rear axle and the vehicle is propelled forward. The concept vehicle incorporates a lever arm that lengthens the snap arm, pulls more string from the axle, and changes the velocity ratio between the mousetrap and the drive wheels.

Although this concept has the positive attribute of being simple and straightforward to construct, the team recognizes that a number of questions are raised, and they also list these in their notebook.

- What should be the length of the lever arm extension and the radius of the spool attached to the drive axle? With a long-enough string, the vehicle would be gradually powered by the mousetrap over the entire 10-m distance. On the other hand, if the string is shorter, the mousetrap will close sooner and the vehicle would coast after being powered only along the first portion of the race course. The team's discussion of this issue prompts the idea for a tapered spool, as sketched in Figure 8.4(c), which would enable the velocity ratio between the mousetrap and the drive axle to change as the mousetrap closes.

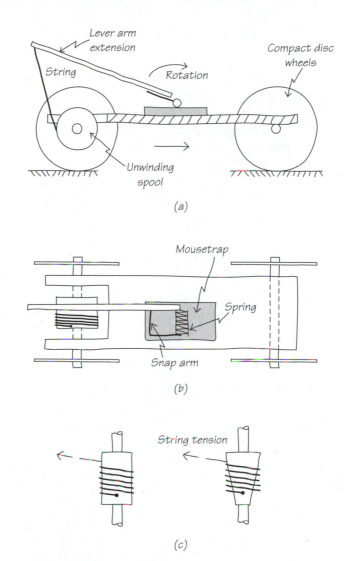

FIGURE 8.4   First concept that is based on a lever arm for pulling and unwrapping string from the drive axle. (a) Side view with two wheels removed for clarity. (b) Top view. (c) Straight and tapered unwinding spools.

- Should the mousetrap be positioned behind, above, or in front of the drive axle? In their concept sketch, the students drew the mousetrap directly between the front and rear wheels. At this stage, however, that placement is arbitrary, and the team has no reason to expect that choice to be better than any other.
- What should be the radius of the wheels? Like the length of the lever arm extension and the radius of the spool, the radius of the drive wheels influences the velocity of the vehicle. The team noted on its sketch that computer compact discs could be used

as the wheels, but the vehicle might post a better race time with different-diameter wheels.

The team records these questions and discussion topics in their notebook, but they leave them for future consideration. At this early point in the brainstorming process, no decisions will be made on dimensions or materials. However, if this concept eventually emerges as a promising candidate, the team will need to resolve these issues before a viable prototype could be constructed.

Second Concept: Compound Geartrain

As their discussions continue, the team next devises the option shown in Figure 8.5 in which a compound geartrain transfers power from the mousetrap to the drive axle. This vehicle has only three wheels, and a portion of the body has been removed to further

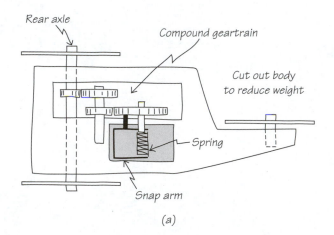

(a)

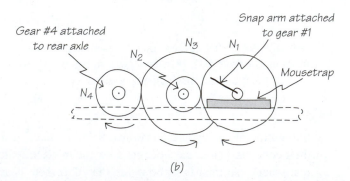

(b)

■ **FIGURE 8.5** Second concept that is based on a compound geartrain between the snap arm of the mousetrap (which rotates through one-half turn) and the drive axle (which is powered over the full 10-m distance). (a) Top view of the vehicle. (b) Layout of the two-stage geartrain.

reduce weight. The concept incorporates a two-stage geartrain, and its velocity ratio is set by the numbers of teeth on the four gears shown in Figure 8.5(b). The team's illustration of a two-stage geartrain is arbitrary; a system with only one stage, or more than two stages, might be preferable. However, the students accept such ambiguity, and they realize that a decision for the geartrain's velocity ratio is not necessary at this early juncture.

During the give-and-take of their meetings, the team identifies additional constraints that are common to their first and second concepts. For instance, the students agree that the vehicle should be designed so that the drive wheels do not spin and slip as the vehicle accelerates. Otherwise, some portion of the limited energy that is available from the mousetrap spring would be wasted. To prevent slippage, weight could be added to the vehicle in order to improve contact between the drive wheels and the ground. On the other hand, a heavier vehicle would travel slower as the potential energy of the mousetrap spring is converted into the vehicle's kinetic energy. As they investigate the project in more detail, the students see that the technical issues at hand are interrelated. Even in the context of this seemingly straightforward exercise, not to mention such challenges as the mechanical engineering profession's top ten achievements (Section 1.3), designers must grapple with competing constraints and specifications.

Third Concept: Sector-Shaped Gear

The team's third concept combines and extends certain ideas that arose during their discussions of the earlier concepts. The design of Figure 8.6 incorporates a geartrain between the mousetrap and the drive wheels, but it enables the vehicle to coast once the trap has closed. The students envision such a vehicle as accelerating quickly over the first few meters of the race course, reaching peak velocity, and then coasting over the remaining distance. In their concept, a sector-shaped gear, instead of a full circular one, is attached to the snap arm of the mousetrap. A small notch at one end of the gear enables the mousetrap to disengage from the simple geartrain once the snap arm has closed, as shown in the Figure 8.6(c). The sector gear serves as the input to the geartrain, and the output gear is directly attached to the front drive axle. The idler gear is included to increase the offset between the mousetrap and the front axle.

With several concepts in mind, the students are now at a point to begin making trade-offs, narrowing down their options, and experimenting with prototypes. In addition, they consider various materials that are available and desirable for construction. Although it is still premature to select specific components, the students use their imaginations to list some options: foam core poster board, balsa and poplar wood, aluminum and brass tubing, threaded rod, plexiglass, ball bearings, oil and graphite lubricants, wire, and epoxy. After addressing some of the technical issues that have been raised, and performing back-of-the-envelope calculations and order-of-magnitude approximations, the team might next decide to build and test a few prototypes before selecting one to be refined in detail.

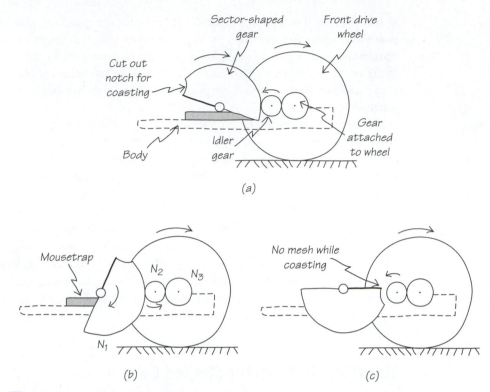

FIGURE 8.6 Third concept that is based on a simple geartrain and a sector-shaped gear. The vehicle is powered over the first portion of the race course and then coasts at top speed over the remaining distance. Side views of the geartrain (a) as the mousetrap begins to close, (b) during the powered phase, and (c) during the coasting phase where the notch disengages the sector-shaped gear from the drive wheels.

8.4 CASE STUDY IN COMPUTER-AIDED DESIGN: NONINVASIVE MEDICAL IMAGING

In Chapter 1, we introduced computer-aided engineering and described its impact as one of mechanical engineering's top ten achievements. Engineers design, analyze, simulate, and manufacture products with the aid of special-purpose computer software that enables those tasks to be performed faster and more accurately than would otherwise be possible. Just as they use equations, drawings, calculators, pencil and paper, and experiments, mechanical engineers apply computer-aided engineering software in their everyday work of solving technical problems. By creating and revising designs electronically and by simulating the performance of those designs in a virtual sense before hardware is ever built, engineers have greater confidence that their products will perform as expected. By way of a case study in this section, we trace the role of computer-aided engineering during the development of a small but critical component of a product that is used during high-resolution medical imaging. The objective of this case study is to highlight for you the manner in which the component was designed through a seamless step-by-step, computer-aided engineering process.

The background of the product at hand involves the technology of magnetic resonance imaging (MRI), which is used by the medical industry to produce images of organs and tissue within the human body. MRI is based upon the principle of nuclear magnetic resonance, a technique developed by scientists to obtain chemical information about the molecular structure of matter. One aspect of MRI is that a liquid chemical called a *contrast agent* can be injected into the patient who is undergoing the examination in order to bring out visible detail in the images of the tissue being examined. For instance, contrast agents increase resolution and enhance the definition between normal and abnormal tissues in the body, as seen in the medical image in Figure 8.7. After the chemical agent is injected, it accumulates in the abnormal tissue, and those areas become brightened in the final image. With such information, a physician or surgeon can develop an improved diagnosis of the patient's medical condition.

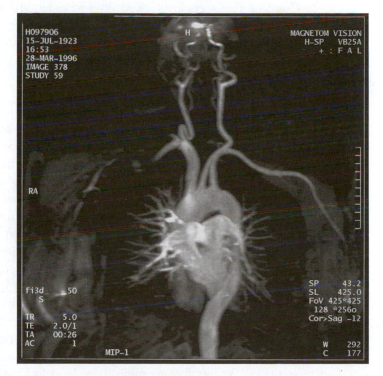

■ **FIGURE 8.7** Noninvasive imaging technologies have revolutionized many aspects of the medical industry, and they provide physicians and surgeons with the information they need to make accurate diagnoses and devise treatment plans.

Source: Reprinted with permission of Medrad, Incorporated.

Contrast agents must be injected in a precise and safe manner. For that reason, automated mechanical syringes under computer control are often used to perform the procedure. The particular system examined in this section comprises two syringes that deliver the contrast agent and a companion saline solution to the patient. The system includes pistons and cylinders as in traditional syringes, but an electronic motor automatically depresses the pistons to precisely inject the chemicals during the MRI session. The syringes are used only once, and then they are disposed. A constraint on the product's design is the method by which the syringes are inserted, held in the automated injection system, and then removed. Our case study in computer-aided engineering involves the connection or interface between the disposable syringe itself and the mechanism that automatically depresses the piston.

Mechanical engineers design the connection between the syringe and the injection system so that a medical technician can quickly remove an empty syringe and install a fresh one. In addition, the connection must be strong enough to securely lock the syringe into place and to neither leak nor break when it is subjected to high pressure during the injection. Engineers designed the syringe and its connection to the injection system through a sequence of steps that draw extensively on computer-aided design tools:

1. **Concept.** Engineers first created a computer-based drawing of each component in the injection system. The cross-sectional view of Figure 8.8(a) illustrates how the syringe interface, cylinder, and piston connect to one another and to the body of the automated injection machine. At the design concept stage, engineers fixed their ideas with an approximate representation of the product, recognizing that many details remained to be resolved.

2. **Detailed design evolution.** As the concept was reviewed and discussed, engineers began incorporating realistic features and tending to details that had, rightly so, not been addressed at the earlier concept stage. The final shape of the syringe interface was established by including all of the geometric features that would be present once the component was ultimately manufactured. The drawing of Figure 8.8(a) was first developed into the three-dimensional solid model shown in Figure 8.9(a). Each detail that would be present in the final component, even the stiffening ribs shown in Figure 8.9(b), was then built into the computer model so as to make it as realistic and representative of finished hardware as possible. Engineers used such drawings to visualize the product and describe its dimensions, shape, and function to others. In addition, the drawings were developed in such a format that other computer-aided engineering tools could directly import the three-dimensional representation, simplifying subsequent stress and manufacturing analyses.

3. **Strength and deformation analyses.** As the syringe is inserted into the automated injection system by the MRI technician, rotated, and snapped into place, the flanges on the syringe interface are subjected to large locking forces that could cause it to crack and break. Because the assembly is used in a precision medical environment, it is critical that engineers design each component to be as reliable as possible. In the next step along the product development process, engineers analyzed the syringe

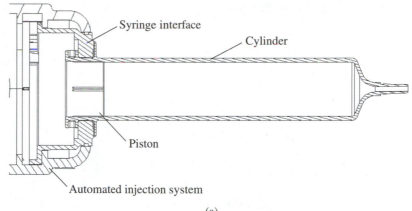

(b)

■ FIGURE 8.8 (a) A computer-generated drawing of the design concept for the syringe. (b) Photograph of the finished product.

Source: Reprinted with permission of Medrad, Incorporated.

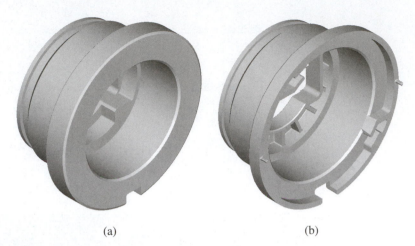

(a)   (b)

■ **FIGURE 8.9** (a) As the design evolved, the computer model was extended to represent the component in three dimensions. (b) The design was refined to reflect all geometric features that would be present once the syringe interface was manufactured.
Source: Reprinted with permission of Medrad, Incorporated.

interface and modified its design so that the flanges would be strong enough for its intended use.

Many computer-aided mechanical engineering tools are compatible with one another, enough so that data files for the dimensions and shape of a component can be transferred between software packages. In that manner, the three-dimensional computer model of Figure 8.9(b) was directly transferred from a drafting software package into one that is used to analyze stress. In a virtual environment, before any parts were actually produced, mechanical engineers simulated how the syringe interface would bend and distort as it is inserted into the injection system (Figure 8.10). If the stress or deformation was predicted to be too large, the engineers would iterate back to the previous step and modify the shape or dimensions until the design had sufficient mechanical strength. As is almost always the case, the process involved several iterations in which the design was repeatedly analyzed, modified, and reanalyzed until the performance requirements were met and the syringe interface would not be expected to break during use.

4. **Manufacturing process.** In the next step, mechanical engineers needed to determine which tools and processes would be used to manufacture the product. Engineering involves designing not only the product but also the techniques that will be used to manufacture it. The engineers in this case decided that the syringe interface would be produced from plastic and that molten material would be injected at high pressure into a mold. Once the plastic cooled and solidified, the mold would be opened and the finished part could be removed. The mechanical engineers therefore needed to design the mold and verify that it would fill with molten plastic as expected.

Figure 8.11 depicts an exploded view of the mold's final design. However, before the mold itself was machined, engineers first used computer-aided engineering tools

■ **FIGURE 8.10**   Engineers performed a computer analysis to simulate the stress within the flanges of the syringe interface.

Source: Reprinted with permission of Medrad, Incorporated.

to analyze and refine it. As shown in Figure 8.12, the injection molding process was simulated as molten plastic flowed into and filled the hollow portions of the mold. In virtual computer simulations, engineers were able to adjust the locations of the injection points, air bleed points, and seams in the mold until the results showed that air bubbles would not become trapped in the mold and that the plastic would

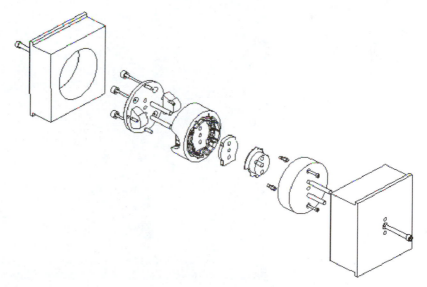

■ **FIGURE 8.11**   Mechanical engineers designed each component of the mold that would be used to manufacture the syringe interface.

Source: Reprinted with permission of Medrad, Incorporated.

Air pockets

■ **FIGURE 8.12**   A computer simulation of molten plastic flowing into and filling the mold in order to identify locations where air bubbles could potentially be trapped.
Source: Reprinted with permission of Medrad, Incorporated.

not cool and solidify before the mold became completely filled (Figure 8.12). If a simulation revealed such problems, the engineers would iterate and change the design of the mold, or the temperature and pressure of the plastic when it is injected until the performance was judged to be satisfactory. Once the design was completed, prototype pieces of the mold were produced for small-scale production. Figure 8.13

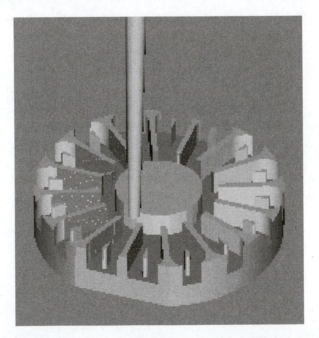

■ **FIGURE 8.13**   A computer simulation of the process for machining one piece of the mold.
Source: Reprinted with permission of Medrad, Incorporated.

depicts a virtual simulation of one mold component as it is machined on a computer-controlled mill.

5. **Design implementation.** Finally, the mechanical engineers prepared detailed technical drawings for the syringe interface and the mold that would be used for its large-scale production (Figure 8.14). Technical reports, test data, and computer analyses were compiled and archived electronically in order to fully document and record the design. In the future, the syringe interface might be modified for use in a new product, and the engineers working on that project would need to review the present design process before they build upon it and develop the next-generation product.

A noteworthy aspect of computer-aided engineering technology is the manner in which each analysis tool can be integrated with the others. For instance, once the three-dimensional solid model of the syringe interface was generated, it was directly imported by the other software tools. Such compatibility greatly simplifies iteration between the design, analysis, and manufacturing stages of product development. This case study highlights what has become known as seamless or paperless computer-aided engineering: A product can be designed, analyzed, prototyped, and manufactured by combining virtual simulation and computer analysis tools throughout the design cycle.

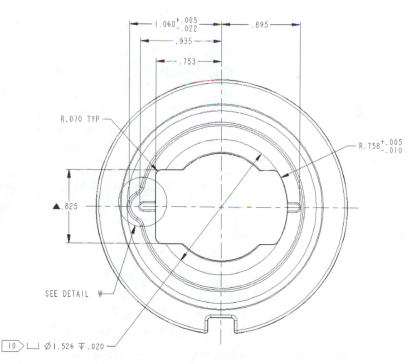

■ FIGURE 8.14   Final verified and documented design of the syringe interface.
Source: Reprinted with permission of Medrad, Incorporated.

8.5 CASE STUDY IN MACHINE DESIGN: THE HYDRA-MATIC TRANSMISSION

Automobile automatic transmissions are an intricate blend of mechanical, electronic, computer, and hydraulic components that operate in concert to produce smooth speed shifts. In this case study, we describe the design of an automatic transmission as an example of a complex machine that is clever, practical, and encountered every day. There are many vehicle-specific types of automatic transmissions, so by way of an introduction in this section, we will discuss the Hydra-Matic transmission, which was the first fully automatic system developed for the consumer market (Figure 8.15). Quite aside from the technical aspects, this particular mechanical engineering technology made a remarkable business contribution to the automotive industry.

The Hydra-Matic transmission was developed by the General Motors Corporation, and it represented a key milestone in the history of the automobile. The transmission was offered on select models of the 1940 Oldsmobile line, and it added less than $60 to the purchase price. The Hydra-Matic was advertised as offering the "glorious new sensation" of driving without the need to operate a clutch pedal. As automobiles were

■ **FIGURE 8.15** Advertisement for the 1940 Oldsmobile highlighting "a car without a clutch pedal."

Source: © 2001 General Motors Corporation. Used with permission of GM Media Archives.

becoming more commonplace and as speeds were increasing, the automatic transmission was further viewed as a safety feature to the extent that the driver did not need to remove a hand from the steering wheel to shift gears. The Hydra-Matic transmission was a commercial success, and within ten years, over 1 million cars with that transmission had been manufactured. Although the Hydra-Matic was originally intended for passenger automobiles, heavier-duty models were adapted during World War II for military armored and amphibious vehicles (Figure 8.16).

Automobile engines operate most efficiently over a limited range of engine speed, neither too low nor too high. The transmission is the means by which vehicle speed can be varied while the engine continues to run within its peak performance range. In the rear-wheel-drive vehicle of Figure 7.20, the transmission is located directly behind the engine. The ridge or bump in the floorboard along the vehicle centerline accommodates the transmission and driveshaft. There are other configurations for a vehicle's drivetrain besides that shown in Figure 7.20, including front-wheel-drive, four-wheel-drive, and all-wheel-drive systems. In some automobiles, the engine is mounted transversely or at the rear of the vehicle. Each configuration is well suited for a particular type of vehicle or niche in the consumer market (Figure 8.17).

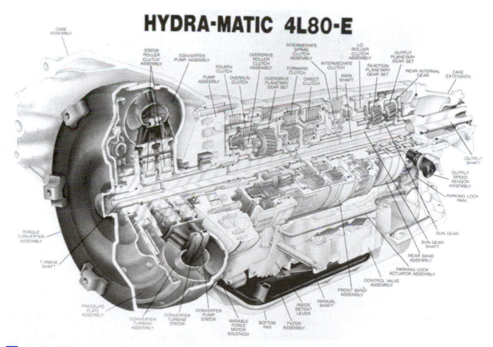

FIGURE 8.16   Cutaway view of the Hydra-Matic transmission.

Source: © 2001 General Motors Corporation. Used with permission of GM Media Archives.

■ **FIGURE 8.17** Computer model depicting the drivetrain in a sport utility vehicle.
Source: Reprinted with permission of Mechanical Dynamics, Incorporated.

Figure 7.20 illustrates the layout of the engine, transmission, and driveshaft in a rear-wheel-drive automobile. In the lowest transmission setting (which is called first gear), the engine crankshaft (the input to the transmission) rotates faster than the driveshaft (the transmission's output). In fourth gear, the Hydra-Matic's highest setting, the transmission has a velocity ratio of 1, and the engine and driveshaft rotate at the same speed. Together with those speed changes, automatic transmissions also modify the torque that is supplied by the engine to the driveshaft according to the ideal geartrain principle $(VR)(TR) = 1$, as discussed in Chapter 7. Of course, automobile transmissions also have a neutral setting, in which the engine is disconnected from the driveshaft, and a setting for reverse.

Modern transmissions incorporate electronic and computer control, and although they are more sophisticated than the Hydra-Matic, they are similar in their basic operating principles. As depicted in Figure 8.18, a fluid coupling serves as the clutch between the engine and the transmission. The Hydra-Matic transmission in Figure 8.18 has four forward speeds, a neutral setting, and one reverse speed. The velocity ratio for each setting is determined by the operation of the three interconnected planetary geartrains, which are called the front, rear, and reverse stages. A hydraulic control unit, which is not shown in Figure 8.18, performs the shifts between transmission settings. Each shift is accomplished by engaging or disengaging the two band brakes, the two clutches, or the reverse lock. The entire transmission comprises a fluid coupling, three

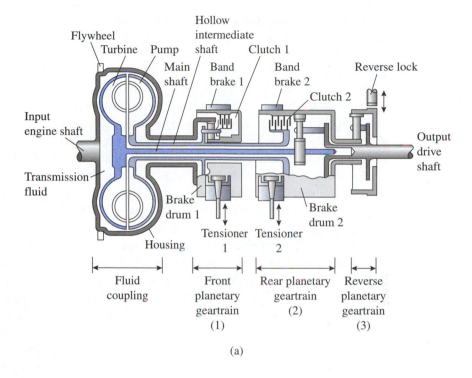

(a)

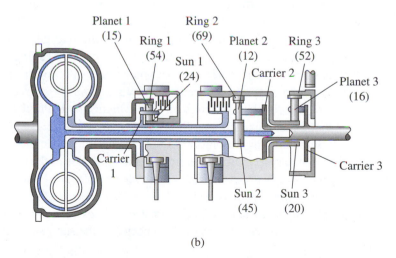

(b)

■ **FIGURE 8.18**   (a) Cross-sectional view of the Hydra-Matic transmission. (b) Sun gears, planet gears, ring gears, and carriers in the front, rear, and reverse planetary geartrains. The tooth numbers are indicated in parentheses.

planetary geartrains, two clutches, two band brakes, a reverse lock, and a hydraulic control system. Those components operate as follows.

Fluid Coupling   This component transfers power from the engine to the transmission without a rigid mechanical connection, a feature that is advantageous for two reasons. First, when the brakes are applied and the transmission is placed in gear, the engine can continue to operate at its idle speed even though the wheels do not rotate. Second, when there is a sudden speed change, the fluid coupling acts as a buffer to isolate other components in the drivetrain from shock. You can view the fluid coupling as being analogous to two household fans that face one another. Imagine that one fan (which we call the pump) is powered up to speed and blows air directly at the (otherwise unpowered) second fan (which we call the turbine). In the ideal situation, the second fan would spin up to the same speed as the first fan and be powered entirely by the air blown on it. In a similar manner, the pump and turbine blades in the fluid coupling of the transmission smoothly transfer power from the engine to the planetary geartrains. The pump and turbine in Figure 8.18(a) are made up of multiple blades, and they are connected to separate shafts placed along the centerline of the transmission. The main shaft is connected to the turbine, and it rotates within the hollow intermediate shaft, which in turn is connected to the pump. The housing in which the pump and turbine rotate is filled with a light oil called *transmission fluid*. As the pump rotates, its blades force oil against the blades on the turbine, causing it to rotate. When the automobile is traveling at a steady speed, the rotation of the turbine on the main shaft follows the rotation of the pump on the hollow intermediate shaft, and they rotate in the same direction and with the same angular velocity.

Planetary Geartrains   The front, rear, and reverse planetary geartrain stages each include a sun gear, planet gears, a ring gear, and a carrier. The tooth numbers for each gear are listed in Figure 8.18(b). The front and rear geartrains set the velocity ratios of the transmission in the first through fourth speed settings, and the reverse stage is used only for the reverse setting. A noteworthy aspect of the transmission is the manner in which the cascade of three planetary geartrains is interconnected through the band brakes and clutches, which in turn are switched by hydraulic control to change speeds. The engine crankshaft supplies power to the transmission by rotating the housing of the fluid coupling. In Figure 8.18(a), the housing is connected to the ring gear of the first planetary geartrain. Therefore, at the input side of the transmission, the automobile engine directly drives the ring gear in the first planetary stage. On the other hand, at the output of the transmission, the driveshaft connects to the carriers on both the rear and reverse planetary stages. The main shaft rotates with the turbine in the fluid coupling, and at its other end, the shaft is connected to the sun gear in the rear planetary stage.

Band Brakes and Reverse Lock   Band brakes in the transmission wrap around drums in the front and rear planetary geartrains. Those brakes are applied or released by the two tensioners shown in Figure 8.18(a), which are activated by the hydraulic control system to effect a speed change. When either brake is applied, it locks the planetary geartrain component to which it is attached and prevents it from rotating. When the band brake in the front stage is applied, for instance, the sun gear is prevented from rotating. Similarly, when the second band brake is applied, the ring gear in the rear planetary stage, and the sun gear in the reverse stage, are both locked stationary. When the reverse lock is applied, the ring gear in the reverse planetary geartrain does not rotate.

Clutches   Like the band brakes and reverse lock, the two clutches in the Hydra-Matic transmission engage or disengage under hydraulic control to change speed settings. The clutches are interleaved disks and plates, and they are used either to connect two rotating components or to disconnect them from one another. When the plates and disks are pressed against one another by a hydraulic piston (not shown in Figure 8.18), the two sides of the clutch rotate at the same speed. If that pressure is instead removed, then no friction or contact between the disks and plates occurs, and the two sides rotate independently. When the clutch in the front planetary geartrain is engaged, the carrier (which is also connected to the hollow intermediate shaft) rotates at the same speed as the sun gear. Likewise, when the clutch in the rear planetary geartrain engages, the hollow intermediate shaft rotates together with the ring gear in the rear stage, and the sun gear in the reverse stage.

Hydraulic Control System   Functioning as a mechanical computer, the hydraulic control system engages and disengages the clutches, band brakes, and reverse lock at the right instants and in the proper combinations to produce smooth shifts in the speed setting. The inputs to the hydraulic control system are (1) the position of the driver's shift lever, which selects either neutral, drive, low range, or reverse settings; (2) the engine speed; and (3) the setting of the driver's accelerator pedal (or engine throttle). The control system in the Hydra-Matic transmission operated entirely by hydraulic pressure, and it contained a complex assemblage of valves, oil passageways, and pistons. The valves moved in response to the inputs of speed and throttle in order to cause the transmission to shift to a new speed setting at the appropriate time. The Hydra-Matic's brake and clutch settings for first through fourth speeds, and reverse, are listed in Table 8.1. In the first setting, for instance, both brakes are applied to the front and rear planetary geartrains, and the transmission produces an overall velocity ratio of 0.273 between the driveshaft and engine. In modern automatic transmissions, the control system incorporates a combination of computer, electronic, hydraulic, and electromechanical components.

We next apply our background in the motion of machinery, and specifically planetary geartrains in Chapter 7, to calculate the velocity ratio between the engine and driveshaft in the first transmission speed setting. We adopt the subscripts r, c, and s to denote ring gears, carriers, and sun gears, and the subscripts 1, 2, and 3 to identify the front, rear, and reverse planetary geartrains. With that terminology and referring to Figure 8.18, $\omega_{r1} = \omega_{\text{engine}}$. Because the carrier in the front geartrain connects to the hollow intermediate shaft, it rotates with the pump in the fluid coupling, and $\omega_{c1} = \omega_{\text{pump}}$.

TABLE 8.1 Brake and Clutch Settings and Overall Transmission Velocity Ratios, $VR = \omega_{\text{driveshaft}}/\omega_{\text{engine}}$

Transmission Speed Setting	Front Planetary Geartrain	Rear Planetary Geartrain	Reverse Planetary Geartrain	Velocity Ratio VR
First	Brake 1	Brake 2	—	0.273
Second	Clutch 1	Brake 2	—	0.395
Third	Brake 1	Clutch 2	—	0.692
Fourth	Clutch 1	Clutch 2	—	1
Reverse	Brake 1	—	Reverse lock	−0.232

When a speed change takes place, or when the transmission is subjected to a sudden torque load, the pump and turbine rotate at different speeds, but at a steady operating condition, $\omega_{\text{turbine}} = \omega_{\text{pump}}$. Because the turbine drives the sun gear of the rear planetary stage, $\omega_{s2} = \omega_{\text{turbine}} = \omega_{\text{pump}} = \omega_{c1}$. The driveshaft rotates at the same speed as the carriers in the rear and reverse geartrains, and so $\omega_{c2} = \omega_{c3} = \omega_{\text{driveshaft}}$.

In examining the first speed setting, our objective is to use Equation 7.19 and calculate the overall velocity ratio $\omega_{\text{driveshaft}}/\omega_{\text{engine}}$. The transmission's three planetary geartrains are similar to the form shown in Figure 7.15, with the tooth numbers listed in Figure 8.18(b). For the front planetary geartrain, $N_{r1} = 54$, $N_{s1} = 24$, and $N_{p1} = 15$, and by using Equation 7.20, the form factor for the front geartrain is $n_1 = 24/15 = 1.6$. For the rear planetary geartrain, the form factor is $n_2 = 45/12 = 3.75$ since $N_{r2} = 69$, $N_{s2} = 45$, and $N_{p2} = 12$.

Referring to Table 8.1, band brakes 1 and 2 are both engaged in the first speed setting, and the transmission can be viewed as taking on the equivalent configuration shown in Figure 8.19. With respect to the front planetary geartrain, $\omega_{s1} = 0$ because brake 1 is engaged, and we have

$$\omega_{r1} = \frac{2(1 + 1.6)}{2 + 1.6}\omega_{c1} = 1.44\omega_{c1} \tag{8.1}$$

after applying Equation 7.19. Since $\omega_{c1} = \omega_{s2}$ and $\omega_{r1} = \omega_{\text{engine}}$, $\omega_{\text{engine}} = 1.44\omega_{s2}$. The engine speed is diminished by a factor of 1.44 as it is fed from the front to the rear planetary geartrain. With brake 2 applied in the second stage, $\omega_{r2} = 0$ and

$$\omega_{s2} = \frac{2(1 + 3.75)}{3.75}\omega_{c2} = 2.53\omega_{c2} \tag{8.2}$$

Since $\omega_{c2} = \omega_{\text{driveshaft}}$, the overall transmission velocity ratio becomes

$$VR = \frac{\omega_{\text{driveshaft}}}{\omega_{\text{engine}}} = \frac{1}{(1.44)(2.53)} = \frac{1}{3.66} = 0.273 \tag{8.3}$$

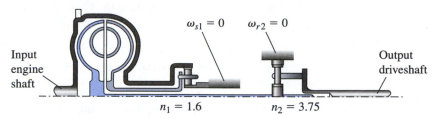

$\omega_{s1} = 0$ $\omega_{r2} = 0$

Input engine shaft

Output driveshaft

$n_1 = 1.6$ $n_2 = 3.75$

■ **FIGURE 8.19** Equivalent configuration of the transmission in the first speed setting where brakes 1 and 2 are engaged. Other elements of the transmission are omitted for clarity.

as listed in the final column of Table 8.1. When the engine operates at 2500 rpm and the transmission is placed in the first speed setting, for instance, the driveshaft will rotate at (2500 rpm)(0.273) = 683 rpm.

8.6 PATENTS IN ENGINEERING

In the process of developing a new product, an engineer or design team may have created novel or breakthrough technology that is important to their company and that they may want to prevent others from using. Patents are an important aspect of the business side of engineering because they provide legal protection to those who invent new technology. Patents are one aspect of intellectual property (a field that also encompasses copyrights, trademarks, and trade secrets), and they are a right to property, analogous to the deed for a building or a parcel of land.

Patents are granted for a new and useful process, machine, manufacture, or composition of matter, or an improvement of them. They are essentially agreements between an inventor and a government. An inventor is granted the legal right to exclude others from making, using, offering for sale, selling, or importing an invention. In exchange, the inventor agrees to disclose and explain the invention to the general public in the written document called a *patent*. A patent is a monopoly on the new technology that expires after a certain number of years, the duration depending on the type of patent that was issued. On the one hand, the inventor provides a service to society by letting others learn about the invention's new technology. On the other, the owner of the patent can license the invention to other individuals or companies who would pay a fee or royalty in return.

It could be argued that the benefits of the patent system have formed the foundation upon which our society has made its technological progress. Patents stimulate corporate research and product development because they provide a financial incentive (the limited monopoly) for innovation. Patents are often highly valuable to small start-up companies because they reduce the other advantages that a larger competing company might have. In short, by being creative, an inventor can use the protection offered by a patent to obtain an advantage over business competitors.

The United States Constitution provides the Congress with the authority to enact patent laws. In fact, Article I, Section 8, of the Constitution states:

Congress shall have the power . . . to promote the progress of science and useful arts, by securing for limited times to authors and inventors the exclusive right to their respective writings and discoveries.

It is interesting to note that in the sequence of powers granted to the government's legislative branch by that portion of the Constitution, the authority to grant patents is mentioned before other (perhaps more well-known) powers of the Congress, such as declaring war and maintaining an army.

There are three types of patents: plant, design, and utility. As the name implies, a plant patent is issued for certain types of asexually reproduced plants, and

Design patent ------------→ it is not commonly encountered by mechanical engineers. A design patent is directed at original ornamental designs. Some examples of products that can be covered by design patents include clothing patterns, bottle shapes, artificial wood grain textures, and toy figures. A design patent is intended to protect an aesthetically appealing product and one that is the result of artistic skill. It does not however, protect the product's so-called functional characteristics. For instance, the shape of an automobile body could be protected by a design patent if it is attractive, pleasing to look at, or gives the vehicle a sporty appearance. However, the functional characteristic of the body, which might include reducing wind drag or offering improved crash protection, would not be protected by a design patent.

Utility patent ------------→ More common in mechanical engineering is the utility patent, which protects the function of an apparatus, process, product, or composition of matter. The utility patent generally contains three main components. The specification is a written description of the purpose, construction, and operation of the invention; the drawings show one or more versions of the invention; and the claims explain in precise language the specific features that the patent protects. The description in the patent must be detailed enough to teach someone else how to practice the invention. As an example, the cover page from a United States patent is shown in Figure 8.20. The page

Patent duration ------------→ includes the patent's title and number, the date when the patent was granted, the names of the inventors, a bibliography of related patents, and an abstract or brief summary of the invention. Utility patents become valid on the date when the patent is granted, and recently issued ones expire 20 years after the date of the application.

The patent application process is a formal one. Engineers normally work with patent attorneys, who conduct a search of existing patents, prepare the application, and interact with the United States Patent and Trademark Office. The application distinguishes what is already known from the new technology that the patent discloses, and it includes a discussion of the invention's background and purpose, drawings, a summary, and one or more claims. A claim describes one aspect of the invention, but a patent can contain one or dozens of them. If it is granted, the patent will generally protect only the technology that is exactly and precisely described by the claims, and so their wording is carefully

US005855257A

United States Patent [19]

Wickert et al.

[11]	Patent Number:	**5,855,257**
[45]	Date of Patent:	**Jan. 5, 1999**

[54] **DAMPER FOR BRAKE NOISE REDUCTION**

[75] Inventors: **Jonathan A. Wickert**, Allison Park; **Adnan Akay**, Sewickley, both of Pa.

[73] Assignee: **Chrysler Corporation**, Auburn Hills, Mich.

[21] Appl. No.: **761,879**

[22] Filed: **Dec. 9, 1996**

[51] **Int. Cl.**^{6} **F16D 65/10**

[52] **U.S. Cl.** .... **188/218 XL; 188/218 A**

[58] **Field of Search** .... 188/18 A, 218 A, 188/218 R, 218 XL; 74/574

[56] **References Cited**

U.S. PATENT DOCUMENTS

1,745,301	1/1930	Johnston .
1,791,495	2/1931	Frey .
1,927,305	9/1933	Campbell .
1,946,101	2/1934	Norton .
2,012,838	8/1935	Tilden .
2,081,605	5/1937	Sinclair .
2,197,583	4/1940	Koeppen et al. .
2,410,195	10/1946	Baselt et al. .
2,506,823	5/1950	Wyant .
2,639,195	5/1953	Bock .
2,702,613	2/1955	Walther, Sr. .
2,764,260	9/1956	Fleischman .
2,897,925	8/1959	Strohm .
2,941,631	6/1960	Fosberry et al. .... 188/218 A
3,250,349	5/1966	Byrnes et al. .
3,286,799	11/1966	Shilton .... 188/218 A
3,292,746	12/1966	Robiette .
3,368,654	2/1968	Wegh et al. .
3,435,925	4/1969	Harrison .
3,934,686	1/1976	Stimson et al. .
4,043,431	8/1977	Ellege .... 188/218 A
4,656,899	4/1987	Contoyonis .
5,004,078	4/1991	Oono et al. .
5,383,539	1/1995	Bair et al. .

FOREIGN PATENT DOCUMENTS

123707	7/1931	Australia .
2275692	1/1976	France .... 188/218 A
58-72735	4/1983	Japan .
63-308234	12/1988	Japan .
141236	9/1984	Rep. of Korea .... 18/218 A
254561	9/1925	United Kingdom .... 188/218 A
708083	10/1952	United Kingdom .
857043	12/1960	United Kingdom .... 188/218 A
934096	8/1963	United Kingdom .... 188/218 A
2181199	4/1987	United Kingdom .
2181802	4/1987	United Kingdom .

Primary Examiner—Robert J. Oberleitner
Assistant Examiner—Chris Schwartz
Attorney, Agent, or Firm—Roland A. Fuller, III

[57] **ABSTRACT**

An apparatus for reducing unwanted brake noise has a ring damper affixed around a periphery of a brake rotor in a disk brake system in a manner that permits relative motion and slippage between the ring damper and the rotor when the rotor vibrates during braking. In a preferred embodiment, the ring damper is disposed in a groove formed in the periphery of the disk and is pre-loaded against the rotor both radially and transversely. The ring damper is held in place by the groove itself and by the interference pre-load or pre-tension between the ring damper and the disk brake rotor.

30 Claims, 4 Drawing Sheets

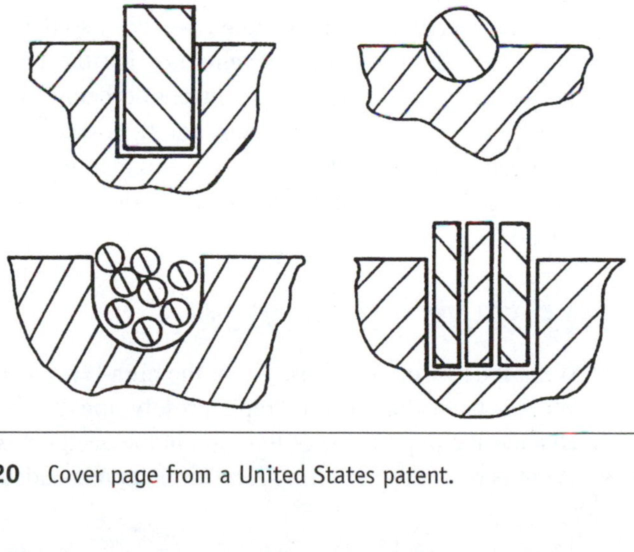

FIGURE 8.20 Cover page from a United States patent.

considered. After the patent is submitted, an examiner from the Patent and Trademark Office reviews the application and informs the inventor as to whether the patent has been accepted. By way of statistics, the patent office received approximately 290,000 patent applications in 2000, and some 180,000 were eventually granted. An applicant for a patent sometimes waits several years for a final decision to be made.

Time restrictions
- - - - - - - - - - - - - - - - ->

There are certain time restrictions involved in obtaining a patent. Most important, perhaps, is the requirement that the application be filed within one year of the inventor having publicly disclosed or used the invention—for instance, by selling or offering to sell it to others, by demonstrating it at an industrial trade show, or by publishing an article on it. In order to document each stage of a product's development, engineers keep records of their design work in logs and notebooks, preferably ones that are bound, numbered, dated, and even witnessed. Drawings, calculations, photographs, test data, and a listing of the dates on which important milestones were reached are important to accurately capture how, when, and by whom the invention was developed. Such records can be critical for obtaining a patent, or for resolving a dispute over who first invented a certain technology.

SUMMARY

The subject of mechanical design has many facets, and in this chapter, we have described them from the conceptual, computer-aided engineering, machine design, and intellectual property viewpoints. At the highest level, engineers apply the procedure of Section 8.2 to systematically reduce a new design problem into a sequence of manageable steps: defining and researching objectives, generating concepts, narrowing down the options, and developing a detailed design. We then examined mechanical design in the context of three case studies in order to learn from the concrete experiences of others. The case studies regarding conceptual design of mousetrap-powered vehicles, computer-aided design of a syringe interface for medical imaging, and machine design of the Hydra-Matic transmission dealt with different aspects of design engineering. In the final section of this chapter, we discussed how patents are used to protect newly-developed technology. Patents provide the inventor with a limited monopoly of the product in exchange for the invention being explained to others. Engineers are creative and conceive new products, and their designs are often patented and become the protected property of a company. Engineering is ultimately a business venture, and you should be aware of that broader context in which mechanical engineering is practiced.

SELF-STUDY AND REVIEW

1. What are the four major stages of the high-level mechanical design procedure?
2. What is the technology of "rapid prototyping"?
3. Discuss the importance of iteration in the design process.
4. What is meant by the term *seamless* computer-aided design?

5. Using Figure 8.18 as a guide, explain the operation of the Hydra-Matic transmission.
6. What are some differences between design and utility patents? How long is a recently issued utility patent valid?
7. How should engineers document and record their design and development efforts?

PROBLEMS

1. Think of an innovative idea that you've had, perhaps for a new time-saving device or something that would improve an existing product. Write a brief description of the application and your concept, and prepare several sketches that explain it.

2. Develop the concept for a new safety feature that can be incorporated in household ladders to prevent falling accidents. Write a brief description of your concept, and prepare several sketches that explain it.

3. Develop the concept for a system to assist disabled persons as they enter and exit a swimming pool. The device is intended to be installed into either new or existing pools. Write a brief description of your concept, and prepare several sketches that explain it.

4. Develop the concept for a packaging system that can prevent the shell of a raw egg from breaking if it is dropped down a flight of stairs. Write a brief description of your concept, and prepare several sketches that explain it.

5. Develop a table that lists advantages and disadvantages of the three design concepts for a mousetrap-powered vehicle. Discuss the trade-offs between front and rear wheel drive, the number of wheels, and the type of drivetrain. Which concept to you think is most viable?

6. Develop and explain another concept for the mousetrap-powered vehicle drive mechanism.

7. The design team decides to use compact discs as wheels because they are lightweight and readily available. However, they must also be aligned and securely attached to the axles. Develop three design concepts for attaching wheels to a 5-mm-diameter axle. Compact discs are 1.25 mm thick, and they have inner and outer diameters of 15 mm and 120 mm.

8. Construct and test a vehicle prototype based on your own concept or one discussed in this section. Prepare drawings of your prototype, and describe its materials and major features.

9. In the second speed setting of the Hydra-Matic transmission, clutch 1 is engaged ($\omega_{s1} = \omega_{c1}$) and brake 2 is applied ($\omega_{r2} = 0$). (a) Show that the sun gear, carrier, and ring gear in the front geartrain rotate at the same speed. (b) Show that $\omega_{engine} = 2.53\omega_{driveshaft}$. (c) Find the driveshaft speed when the engine operates at 2500 rpm.

10. The shift from the second to third speed settings in the original Hydra-Matic transmission involved the simultaneous engagement or disengagement of both band brakes and both clutches. As a result, the shift was perceived by drivers as being rougher than the other speed changes. Calculate the transmission velocity ratio in the third speed setting, where band brake 1 and clutch 2 are engaged, and find the speed of the driveshaft when the engine operates at 2500 rpm.

11. In the fourth speed setting of the Hydra-Matic transmission, clutches 1 and 2 are engaged. Determine the transmission velocity ratio, and find the speed of the driveshaft when the engine operates at 2500 rpm.

12. In the reverse setting of the Hydra-Matic transmission, band brake 1 and the reverse

lock are engaged. Determine the transmission velocity ratio, and find the speed of the driveshaft when the engine operates at 2500 rpm.

13. Explore a patent search engine that is available on the Internet. By using a keyword search for a subject or an inventor's name, find a patent related to mechanical engineering that seems interesting to you. Prepare a one-page description of the patent's application area and its new technology. Explain how you found this patent, and why you think it is interesting. Submit the cover page of the patent and the text of the first claim. The write-up should be prepared on a word processor.

REFERENCE

"The Hydra-Matic, 1939 Thru 1956," *Lubrication*, The Texas Company, Texaco Petroleum Products, February, 1956, pp. 13–32.

A Greek Alphabet

Name	Uppercase	Lowercase	Alternative Symbol
Alpha	A	α	
Beta	B	β	
Gamma	Γ	γ	
Delta	Δ	δ	∂
Epsilon	E	ε	ϵ
Zeta	Z	ζ	
Eta	H	η	
Theta	Θ	θ	ϑ
Iota	I	ι	
Kappa	K	κ	
Lambda	Λ	λ	
Mu	M	μ	
Nu	N	ν	
Xi	Ξ	ξ	
Omicron	O	o	
Pi	Π	π	
Rho	P	ρ	
Sigma	Σ	σ	ς
Tau	T	τ	
Upsilon	Υ	υ	
Phi	Φ	ϕ	φ
Chi	X	χ	
Psi	Ψ	ψ	
Omega	Ω	ω	

B   Review of Mathematical Equations

B.1   TRIGONOMETRIC IDENTITIES

$$\sin^2 \theta + \cos^2 \theta = 1 \tag{B.1}$$

$$\sin^2 \theta = \frac{1}{2}\left(1 - \cos 2\theta\right) \tag{B.2}$$

$$\cos^2 \theta = \frac{1}{2}\left(1 + \cos 2\theta\right) \tag{B.3}$$

$$\sin 2\theta = 2\sin\theta\cos\theta \tag{B.4}$$

$$\cos 2\theta = \cos^2\theta - \sin^2\theta \tag{B.5}$$

$$\sin\left(\theta_1 \pm \theta_2\right) = \sin\theta_1\cos\theta_2 \pm \cos\theta_1\sin\theta_2 \tag{B.6}$$

$$\cos\left(\theta_1 \pm \theta_2\right) = \cos\theta_1\cos\theta_2 \mp \sin\theta_1\sin\theta_2 \tag{B.7}$$

B.2   RIGHT TRIANGLES (Figure B.1)

$$\phi + \theta = \frac{\pi}{2}\ \text{rad} = 90° \tag{B.8}$$

$$x = z\cos\theta = z\sin\phi \tag{B.9}$$

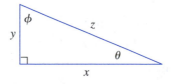

FIGURE B.1   Right triangle with side lengths x, y, z and interior angles ϕ and θ.

$$y = z \sin \theta = z \cos \phi \tag{B.10}$$

$$\tan \theta = \frac{y}{x} \tag{B.11}$$

$$\tan \phi = \frac{x}{y} \tag{B.12}$$

$$x^2 + y^2 = z^2 \tag{B.13}$$

B.3 OBLIQUE TRIANGLES (Figure B.2)

- Area

$$\frac{1}{2}ab \sin C \tag{B.14}$$

- Law of sines

$$\frac{a}{\sin A} = \frac{b}{\sin B} = \frac{c}{\sin C} \tag{B.15}$$

- Law of cosines

$$c^2 = a^2 + b^2 - 2ab \cos C \tag{B.16}$$

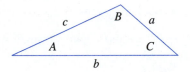

■ **FIGURE B.2**   Oblique triangle with side lengths *a*, *b*, *c* and opposing interior angles *A*, *B*, and *C*.

B.4 DIFFERENTIATION

- Chain rule for $f[\theta(t)]$

$$\frac{df}{dt} = \left(\frac{df}{d\theta}\right)\left(\frac{d\theta}{dt}\right) \tag{B.17}$$

- Sine function

$$\frac{d(\sin \theta)}{d\theta} = \cos \theta \tag{B.18}$$

- Cosine function

$$\frac{d(\cos\theta)}{d\theta} = -\sin\theta \tag{B.19}$$

- Algebraic forms

$$\frac{d(x^2)}{dx} = 2x \tag{B.20}$$

$$\frac{d(\sqrt{1-x^2})}{dx} = \frac{-x}{\sqrt{1-x^2}} \tag{B.21}$$

C Planetary Geartrains

In this Appendix, we derive Equation 7.19, which relates angular velocities of the sun gear (ω_s), carrier (ω_c), and ring gear (ω_r) in a planetary geartrain. Referring to Figure C.1, the pitch radii of the sun and planet gears are represented by r_s and r_p. The carrier's length is $r_c = r_s + r_p$, and the pitch radius of the ring gear is $r_r = r_s + 2r_p$. We conceptually disassemble the geartrain into the components shown in Figure C.2. Our intention is to analyze the rotation of each element individually, and then combine those results to relate shaft speeds.

The sun gear, ring gear, and carrier rotate at different speeds about the center of the geartrain, labeled as point A in Figures C.2 and C.3. The velocity of either point B, C, or D is therefore the product of a rotational speed and a radius, and each is determined by applying Equation 7.1. Taking the sun gear, carrier, and ring gear in sequence, their velocities shown in Figure C.2 are as follows:

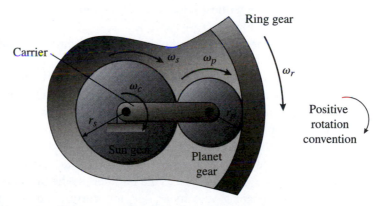

FIGURE C.1   Configuration of a planetary geartrain showing the pitch radii of the sun and planet gears, and the positive sign convention for each rotation.

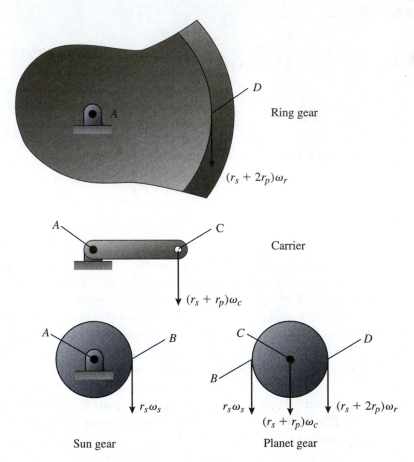

FIGURE C.2 Velocities of the pivot and mesh points *A*, *B*, *C*, and *D*.

Sun gear: The velocity of point *B* on the pitch circle of the sun gear is $r_s \omega_s$.

Carrier: The outermost point *C* of the carrier, which is also the center of the planet gear, has velocity $(r_s + r_p)\omega_c$.

Ring gear: The velocity of point *D* on the pitch circle of the ring gear is $(r_s + 2r_p)\omega_r$.

We next consider the planet gear. As shown in Figures C.3 and C.4, points *B* and *D* on the planet gear contact corresponding points on the sun and ring gears. We view the motion of the planet gear as the combination of two different effects, motivated by the literal "planet" analogy. First, the planet gear rotates about its center, just as the Earth rotates about its center once each day. For that component of motion, a point on the pitch circle of the planet gear has velocity $r_p \omega_p$. Second, the planet gear orbits the center *A* of the geartrain, just as the Earth orbits the Sun once each year. The center *C* of the planet gear has velocity $(r_s + r_p)\omega_c$, corresponding to this orbital component of motion. The velocities of points *B* and *D* on the planet gear are calculated by combining these two effects, a process illustrated in Figure C.4.

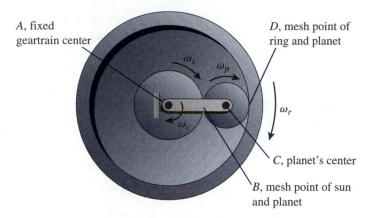

A, fixed geartrain center

D, mesh point of ring and planet

ω_r

C, planet's center

B, mesh point of sun and planet

■ **FIGURE C.3** Definition of the pivot and mesh points A, B, C, and D.

The contact points on the planet gear and sun gear have identical velocities where they mesh, so we have

$$r_s\omega_s = (r_s + r_p)\omega_c - r_p\omega_p \tag{C.1}$$

which is rearranged for the speed of the planet gear as

$$\omega_p = \frac{(r_s + r_p)\omega_c - r_s\omega_s}{r_p} \tag{C.2}$$

Likewise, by matching velocities of the planet gear and ring gear at point D,

$$(r_s + 2r_p)\omega_r = (r_s + r_p)\omega_c + r_p\omega_p \tag{C.3}$$

After substituting for ω_p from Equation C.2, we obtain

$$(r_s + 2r_p)\omega_r = 2(r_s + r_p)\omega_c - r_s\omega_s \tag{C.4}$$

planet gear's motion   =   translation with point C +   rotation about point C

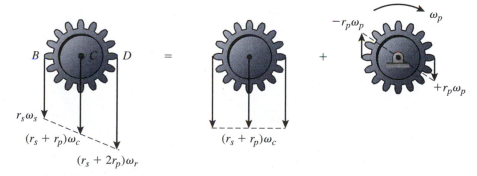

■ **FIGURE C.4** The combination of the two motion components of the planet gear.

an expression that relates speeds at the three connection points of the geartrain: the ring gear (ω_r), the carrier (ω_c), and the sun gear (ω_s).

Finally, we place Equation C.4 into standard form based on the number of teeth for each gear instead of the pitch radius. Because the gears must have the same diametral pitch in order for them to mesh properly, a radius ratio is equivalent to a ratio of teeth numbers. We denote the numbers of teeth on the sun, planet, and ring gears by N_s, N_p, and N_r, and define the form factor

$$n = \frac{r_s}{r_p} = \frac{N_s}{N_p} \tag{C.5}$$

Equation C.4 for the shaft speeds then becomes

$$(2 + n)\omega_r + n\omega_s - 2(1 + n)\omega_c = 0 \tag{C.6}$$

which relates the speeds of the ring gear, carrier, and sun gear.

Index

0.2% offset method, 164

Aerodynamics, 132
Airfoil, 134
Angle of attack, 134
Angular velocity, 235

Balanced geartrain, 249
Ball bearings, 64
Band saw, 89
Base circle, 75
Base units, 39
Bernoulli's equation, 199
Bevel gears, 78
Boiler, 219
Bottom dead center, 213
British thermal unit, 186
Brittle material, 155
Buoyancy force, 122

Cam, 212
Camber, 136
Career ladder, 248
Carnot cycle, 206
Carnot efficiency, 206
Carrier, 248
Celsius, 206
Chain drive, 87
Chain rule, 257
Characteristic length, 127
Chord area, 134

Clerk engine cycle, 216
CNC mill, 94
Communication skills, 29
Compound geartrain, 244
Compression ratio, 213
Computer-aided engineering, 21
Condenser, 220
Conduction, 191
Convection, 194
Core courses, 30
Crank ratio, 258
Creativity, 268
Crossed helical gear, 81

Degrees Kelvin, 206
Degrees Rankine, 206
Derived units, 39
Design notebook, 273
Design patent, 293
Diametral pitch, 74
Differential, 253
Dimensionless number, 48
Double shear, 167
Drag coefficient, 129
Drill press, 88
Ductile material, 154

Efficiency, 205
Elastic behavior, 150
Elastic limit, 158
Elastic modulus, 156

307

CREDITS

This page constitutes an extension of the copyright page. We have made every effort to trace the ownership of all copyrighted material and to secure permission from copyright holders. In the event of any question arising as to the use of any material, we will be pleased to make the necessary corrections in future printings. Thanks are due to the following authors, publishers, and agents for permission to use the material indicated.

Chapter 1

2: Reprinted courtesy of Caterpillar, Incorporated; **3:** Reprinted with permission of NASA; **5:** Reprinted with permission of NASA; **8:** Reprinted with permission of FANUC Robotics North America, Incorporated; **9:** (All) Reprinted with permission of Niagara Gear Corporation, Boston Gear Corporation, and W. M. Berg, Incorporated; **10:** Reprinted with permission of Mechanical Dynamics, Incorporated; **12:** © 2001 General Motors Corporation. Used with permission of GM Media Archives; **13:** Reprinted with permission of NASA; **14:** Reprinted with permission of NASA; **15:** Reprinted with permission of Westinghouse Electric Company; **16:** Reprinted with permission of Enron Wind; **17:** Reprinted with permission of the National Robotics Engineering Consortium; **18:** Reprinted with permission of Lockheed-Martin; **19:** Reprinted with permission of NASA; **20:** (Top) Reprinted with permission of Intel Corporation; **20:** (Bottom) Reprinted with permission of G. Fedder, Carnegie Mellon University; **22:** Reprinted with permission of NASA; **23:** (Top) Reprinted with permission of NASA; **23:** (Bottom) Reprinted with permission of Fluent, Incorporated; **24:** (All) Reprinted with permission of Algor Corporation; **25:** Reprinted with permission of Velocity11; **26:** Reprinted with permission of Fluent, Incorporated

Chapter 2

38: Reprinted with permission of the Winnipeg Free Press; **51:** Reprinted with permission of NASA

Chapter 3

62: Reprinted with permission of Niagara Gear Corporation; **64, 65, 66:** Courtesy of the author; **67:** (Top) Courtesy of the author; **67:** (Bottom) Reprinted with permission of The Timken Company; **68:** Reprinted with permission of The Timken Company; **69:** (Top) Courtesy of the author; **69:** (Bottom) Reprinted with permission of The Timken Company; **71:** Reprinted with permission of W. M. Berg, Incorporated; **73:** (Top) Courtesy of the author; **73:** (Bottom) Reprinted with permission of Boston Gear; **77:** Reprinted with permission of Boston Gear; **78:** Reprinted with permission of Boston Gear; **79:** (All) Reprinted with permission of Boston Gear; **80:** (All) Courtesy of the author; **81:** Reprinted with permission of Boston Gear; **82:** Reprinted with permission of Boston Gear; **83:** (All) Reprinted with permission of Boston Gear; **84:** © General Motors Corporation. Used with permission of GM Media Archives; **86:** (All) Reprinted with permission of W. M. Berg, Incorporated; **87:** Courtesy of the author; **88:** © 2001 General Motors Corporation. Used with permission of GM Media Archives; **89, 90, 91, 92:** Courtesy of the author; **93:** (Top) Reprinted with permission of W. M. Berg, Incorporated; **93:** (Bottom) Courtesy of the author

Chapter 4

97: Reprinted with permission of Mechanical Dynamics, Incorporated, and by Caterpillar, Incorporated; **119:** Reprinted with permission of NASA; **120:** Reprinted with permission of NASA; **121:** Reprinted with permission of NASA; **131:** Reprinted with permission of Fluent, Incorporated; **132:** Reprinted with permission of Lockheed-Martin; **133:** Reprinted with permission of NASA; **134:** Reprinted with permission of Fluent, Incorporated

Chapter 5

149: Courtesy of the author; **160:** Reprinted with permission of MTS Systems Corporation; **172:** Reprinted with permission of Algor Corporation

Chapter 6

181: Reprinted courtesy of Caterpillar, Incorporated; **196:** Reprinted with permission of Mechanical Dynamics, Incorporated; **197:** Reprinted with permission of the Bureau of Reclamation. United States Department of the Interior; **198:** Reprinted with permission of the Bureau of Reclamation, United States Department of the Interior; **202:** Reprinted with permission of Fluent, Incorporated; **210:** © 2001 General Motors Corporation. Used with permission of GM Media Archives; **213:** Reprinted with permission of Mechanical Dynamics, Incorporated; **222:** (Top) Reprinted with permission of Fluent, Incorporated; **222:** (Bottom) Reprinted with permission of Westinghouse Electric Company; **223:** Reprinted with permission of Westinghouse Electric Company; **224:** Reprinted with permission of Westinghouse Electric Company; **225:** (Top) Reprinted with permission of GE Aircraft Engines; **225:** (Bottom) Reprinted with permission of Fluent, Incorporated; **228:** Courtesy of the author

Chapter 7

234: (Top) Reprinted with permission of NASA; **234:** (Bottom) Reprinted with permission of FANUC Robotics North America, Incorporated; **236:** Reprinted with permission of NASA; **240:** Courtesy of the author; **248:** Reprinted with permission of NASA Glenn Research Center; **250:** Reprinted with permission of The Timken Company

Chapter 8

269: Reprinted with permission of Stratasys, Incorporated; **270:** Reprinted with permission of Stratasys, Incorporated; **277:** Reprinted with permission of Medrad, Incorporated; **279:** (All) Reprinted with permission of Medrad, Incorporated; **280:** Reprinted with permission of Medrad, Incorporated; **281:** (All) Reprinted with permission of Medrad, Incorporated; **282:** (All) Reprinted with permission of Medrad, Incorporated; **283:** Reprinted with permission of Medrad, Incorporated; **284:** © 2001 General Motors Corporation. Used with permission of GM Media Archives; **285:** © 2001 General Motors Corporation. Used with permission of GM Media Archives; **286:** Reprinted with permission of Mechanical Dynamics, Incorporated